SPATIOTEMPORAL MODELS OF POPULATION AND COMMUNITY DYNAMICS

Population and Community Biology Series

Principal Editor

M.B. Usher
Chief Scientist, Scottish Natural Heritage, UK

Editors

D.L. DeAngelis
Department of Biology, University of Florida, USA and **B.F.J. Manly** *Director, Centre for Applications of Statistics and Mathematics, University of Otago, New Zealand*

Population and community biology is central to the science of ecology. This series of books explores many facets of population biology and the processes that determine the structure and dynamics of communities. Although individual authors have freedom to develop their subjects in their own way, these books are scientifically rigorous and generally adopt a quantitative approach.

Already published

4. **The Statistics of Natural Selection**
   B.F.J. Manly (1985) 484pp. Hb/Pb. Hb out of print.

5. **Multivariate Analysis of Ecological Communities**
   P. Digby and R. Kempton (1987) 206pp. Hb/Pb. Hb out of print.

6. **Competition**
   P. Keddy (1989) 202pp. Hb/Pb. Hb out of print.

7. **Stage-Structured Populations: Sampling, Analysis and Simulation**
   B.F.J. Manly (1990) 200pp. Hb. Out of print.

8. **Habitat Structure: The Physical Arrangement of Objects in Space**
   S.S. Bell, E.D. McCoy and H.R. Mushinsky (1991, reprint 1994) xiv+438pp. Hb. Out of print.

9. **Dynamics of Nutrient Cycling and Food Webs**
   D.L. DeAngelis (1992) xv+270pp. Pb. ISBN 0 412 29840 6.

10. **Analytical Population Dynamics**
    T. Royama (1992 Hb, reprinted in PB 1996) xvi+371pp. ISBN 0 412 24320 2 (Hb) and 0 412 75570 X (Pb).

11. **Plant Succession: Theory and Prediction**
    D.C. Glenn-Lewin, R.K. Peet and T.T. Veblen (1992) 361pp. Hb. ISBN 0 412 26900 7.

12. **Risk Assessment in Conservation Biology**
    M.A. Burgman, S. Ferson and R. Akcakaya (1993) ix+314pp. Hb. ISBN 0 412 35030 0.

13. **Rarity**
    K. Gaston (1994) x+205pp. Hb/Pb. Hb out of print. ISBN 0 412 47510 3 (Pb).

14. **Fire and Plants**
    W.J. Bond and B.W. van Wilgen (1996) xviii+263pp. Hb. ISBN 0 412 47540 5.

15. **Biological Invasions**
    M. Williamson (1996) xii+244pp. Hb/Pb. ISBN 0 412 31170 4 (Hb) and 0 412 59190 1 (Pb).

16. **Regulation and Stabilization: Paradigms in Population Ecology**
    P.J. den Boer and J. Reddingius (1996) xiii+397pp. Hb. ISBN 0 412 57540 X.

17. **Biology of Rarity**
    W. Kunin and K. Gaston (eds) (1997) xiv+280pp. Hb. ISBN 0 412 63380 9.

18. **Structured-Population Models in Marine, Terrestrial, and Freshwater Systems**
    S. Tuljapurkar and H. Caswell (eds) (1997) xii+643pp. Pb. ISBN 0 412 07271 8.

19. **Resource Competition**
    J.P. Grover (1997) x+342pp. Hb. ISBN 0 412 74930 0.

20. **Individual Behavior and Community Dynamics**
    J. Fryxell and P. Lundberg (1998) ISBN 0 412 9940 11 (Hb) and 0 412 99411 9 (Pb).

21. **Spatiotemporal Models of Population and Community Dynamics**
    T. Czárán (1998) Hb. ISBN 0 412 57550 7.

22. **Ecology of Shallow Lakes**
    M. Scheffer (1998) Hb. ISBN 0 412 74920 3.

Forthcoming

Chaos in Real Data: Analysis of Non LinearDynamics from Short Ecological Time Series
J. Perry, D.R. Morse, R.H. Smith and I.P. Woiwood (eds)

Container Habitats and Food Webs: The Natural History and Ecology of Phytotelmata
R. Kitching

Structure and Dynamics of Fungal Populations
J.J. Worrall

Dynamics of Coral Reef Communities
R.H. Karlson

Ecology of Natural Disturbance in Plant Communities
P. Attiwill

Trophic Cascades in Nature
L. Oksanen

Spatiotemporal Models of Population and Community Dynamics

Tamás Czárán
Ecological Modelling Research Group,
Hungarian Academy of Sciences, and
Department of Plant Taxonomy and Ecology, Eötvös University,
H-1083 Budapest, Ludorika Tér 2, Hungary

CHAPMAN & HALL
London · Weinheim · New York · Tokyo · Melbourne · Madras

Published by Chapman & Hall, 2–6 Boundary Row, London SE1 8HN

Chapman & Hall, 2–6 Boundary Row, London SE1 8HN, UK

Chapman & Hall GmbH, Pappelallee 3, 69469 Weinheim, Germany

Chapman & Hall USA, 115 Fifth Avenue, New York, NY 10003, USA

Chapman & Hall Japan, ITP-Japan, Kyowa Building, 3F, 2-2-1 Hirakawacho, Chiyoda-ku, Tokyo 102, Japan

Chapman & Hall Australia, 102 Dodds Street, South Melbourne, Victoria 3205, Australia

Chapman & Hall India, R. Seshadri, 32 Second Main Road, CIT East, Madras 600 035, India

First edition 1998

© 1998 Támas Czárán

Typeset in 10/12 pt Times by Best-Set Typesetters Ltd, Hong Kong

Printed in Great Britain by St Edmundsbury Press

ISBN 0 412 57550 7

Apart from any fair dealing for the purposes of research or private study, or criticism or review, as permitted under the UK Copyright Designs and Patents Act, 1988, this publication may not be reproduced, stored, or transmitted, in any form or by any means, without the prior permission in writing of the publishers, or in the case of reprographic reproduction only in accordance with the terms of the licences issued by the Copyright Licensing Agency in the UK, or in accordance with the terms of licences issued by the appropriate Reproduction Rights Organization outside the UK. Enquiries concerning reproduction outside the terms stated here should be sent to the publishers at the London address printed on this page.
    The publisher makes no representation, express or implied, with regard to the accuracy of the information contained in this book and cannot accept any legal responsibility or liability for any errors or omissions that may be made.

A catalogue record for this book is available from the British Library

Library of Congress Catalog Card Number: 97–68492

∞ Printed on permanent acid-free text paper, manufactured in accordance with ANSI/NISO Z39.48-1992 and ANSI/NISO Z39.48-1984 (Permanence of Paper).

To the memory of Pál Juhász-Nagy

Contents

Preface xi
Book structure xiv
Acknowledgements xvi

1 Introduction 1
1.1 The simplifying assumptions of classical population dynamics 1
1.2 Spatial extensions of the classical approach 4
1.3 The object-based approach: neighbourhood modelling 5
1.4 The role of spatiotemporal models in ecology 10

Part One Spatial Mass-interaction Models

2 Reaction–diffusion models of population growth and dispersion 13
2.1 Introduction 13
2.2 Random walk approximations to diffusion 15
    2.2.1 Random walk in discrete time and space 15
    2.2.2 Random walk in continuous time and space 17
2.3 The diffusion equation 22
    2.3.1 The flow balance approximation to diffusion 24
    2.3.2 Density flow 26
    2.3.3 Diffusion in 2D and 3D spaces 27
    2.3.4 Methodological overview: initial and boundary conditions, solutions, stability analyses and numerical approximations of PDE models 29
    2.3.5 Solutions of the diffusion equation 34
    2.3.6 Density-dependent diffusion with biased random walk 37
2.4 Advection 38
    2.4.1 Constant rate advection with step probability adjustment 39
    2.4.2 Constant rate advection with step length adjustment 40
    2.4.3 Advection induced by a milieu gradient 41
2.5 Reaction I: population growth in diffusive systems 42
    2.5.1 Constant rate growth and dispersion 43
    2.5.2 The critical habitat size problem 45
    2.5.3 Density-dependent growth and dispersion 46

viii Contents

2.6	Reaction II: species interactions in diffusive systems	49
	2.6.1 The Lotka–Volterra diffusion model	49
	2.6.2 Diffusive instability in models of interacting species	52
	2.6.3 Competitive coexistence through habitat-partitioning	55
2.7	Summary	57

3 Population dynamics in patchy environments — 59

3.1	Introduction	59
3.2	The patch-abundance approach	61
	3.2.1 Basic assumptions of patch-abundance models	61
	3.2.2 The general model	62
	3.2.3 The problem of state variable choice	63
3.3	Competition and mutualism in dispersing island populations	63
	3.3.1 The multispecies multipatch Lotka–Volterra model	68
	3.3.2 Persistence and coexistence conditions	69
	3.3.3 Single species persistence: source–sink dynamics	73
	3.3.4 Habitat fragmentation effects	73
	3.3.5 Spatial pattern and competitive coexistence	74
	3.3.6 Single species resilience and risk spreading	76
3.4	Predation in patchy habitats	78
	3.4.1 Diffusive coupling of identical predator–prey patches	82
	3.4.2 Dispersal asymmetry and stability in a two-patch Lotka–Volterra model	86
	3.4.3 Aggregation and stability in a two-patch environment	88
	3.4.4 The effects of predator mobility and delayed functional response	94
3.5	Chaotic dynamics of single-species systems in patchy environments	95
	3.5.1 Migration against chaos: dispersion and stability in coupled maps	96
	3.5.2 Coupled map lattices: the multipatch extension of the coupled logistic model	99
	3.5.3 Self-organized criticality defeats chaos in a coupled map lattice	104
3.6	Summary	109

4 Spatially implicit patch models: metapopulations and aggregated interactions — 111

4.1	Introduction	111
4.2	Metapopulations and metacommunities	112
	4.2.1 Colonization–extinction equilibrium in the basic model	113

	4.2.2	The ghost of within-patch dynamics returns: the asynchronous age-structured model	114
	4.2.3	Metapopulations with synchronous local dynamics	117
	4.2.4	Rescue effect due to spatial heterogeneity	120
	4.2.5	Patch size and quality effects: phenomenological model	122
	4.2.6	Patch size and quality effects: mechanistic model	124
	4.2.7	Multistate metapopulation models	125
	4.2.8	Interacting metapopulations: the structure of metacommunity models	130
	4.2.9	Competitive metacommunities: continuous time models	131
	4.2.10	Discrete time metacommunity models	134
	4.2.11	Comparing patch-abundance and patch-occupancy models	136
	4.2.12	Connection to island biogeography: incidence function models	137
4.3	Aggregation models of species interactions		140
	4.3.1	The non-spatial reference: the Nicholson–Bailey model	141
	4.3.2	Spatial heterogeneity included: aggregation of encounters in a patchy host distribution	144
	4.3.3	Spatially undetermined aggregation of interactions	149
4.4	Summary		151

Part Two Neighbourhood Models of Population Interactions

5	**Site-based neighbourhood models**		**155**
5.1	Introduction		155
5.2	Interacting particle systems and cellular automata		156
	5.2.1	The structure of interacting particle system models	156
	5.2.2	Mean-field approximations to interacting particle systems	159
	5.2.3	Aspects of complexity in interacting particle systems	163
5.3	Interacting particle systems and cellular automata in ecology		164
	5.3.1	Discrete individuality and dynamical coexistence	165
	5.3.2	Interacting particle system models of competing metapopulations with temporary and permanent habitat destruction	172
	5.3.3	The temporal refuge effect in one-sided competition; an example for a configuration-field approximation	174

	5.3.4	The role of mesoscale patterns in the dynamics of predator–prey cellular automata	182
	5.3.5	Plant competition along an environmental gradient	185
	5.3.6	Plant competition in a fractal environment	188
	5.3.7	The effect of clonal integration on plant competition for mosaic habitat space	195
	5.3.8	Percolation models of spreading populations, epidemics and forest fires	198
5.4	Summary		199

6 Individual-based neighbourhood models   202
6.1 Introduction   202
6.2 Tessellation models   203
    6.2.1 Predicting plant performance from tessellation parameters: the Voronoi assignment model   206
    6.2.2 An interpretation of the self-thinning rule on the individual level   207
    6.2.3 Tessellation models of territory establishment   210
    6.2.4 Towards tessellation dynamics: linking tessellations to demography   212
    6.2.5 Towards multispecies tessellation dynamics: weighted tessellations   213
6.3 Distance models   218
    6.3.1 Fixed radius neighbourhood models   220
    6.3.2 Zone of influence models   231
    6.3.3 Ecological field models   239
6.4 Summary   242

7 Epilogue   244

Appendix A   246
Appendix B   248
Appendix C   250
References   251
Index   270

Preface

The mutual dependence of spatial structure and temporal dynamics is a slumberous common occurrence in recent ecology textbooks. This being the case, the lack of a broad interface between structural and dynamical studies in ecology – which has been, and partly remains, a serious problem of the discipline – might seem puzzling. The overwhelming majority of structural research has been directed towards revealing the spatial pattern and the topographical–topological relations of co-occurring populations by statistical means, while most dynamical studies omitted spatial pattern altogether, allocating considerable effort into the purely temporal description of population processes.

The structural–dynamical dichotomy originated – at least in part – from differences in the attitudes and the methodologies of plant and animal ecology. Taking the Humboldtian viewpoint, most plant ecologists were primarily motivated by the need to discover the structural rules behind the seemingly static patterns of vegetation on many different spatial scales, whereas animal ecology inquired into the more obvious temporal changes of abundance in animal populations and communities, which do not usually seem to preserve a static spatial structure for a long time. The former attitude gave rise to the **structural–static–statistic** discipline of biosociology (coenology), whereas the latter resulted in the development of the classical, **spatially astructural–dynamic–analytic** approach of population dynamics, following in the pioneering steps of Malthusian demography.

Ever since the Humboldtian start, biosociology has remained closely tied to field practice, theoretical efforts directed mostly towards the related statistical methodologies. The interest of biosociologists has been mainly focused on the **spatial** structure of an existing population, or that of a real community. By now, sophisticated oligo- and multivariate statistical methods aid the discovery of spatial patterns in much detail; however, despite repeated contemporary attempts to relate community structure to community dynamics on a statistical basis, these methods alone do not seem sufficient to reveal specific dynamical mechanisms producing specific spatial patterns.

Biosociology does not explain how and why a certain pattern develops in a community, but it can provide a detailed phenomenological description of the spatial pattern in statistical terms, which may give hints on what kind of dynamical rules might have acted in producing a certain structure.

On the other hand, amidst the exciting endeavour to discover and explain mechanisms leading to different **temporal** abundance patterns of populations and communities, classical population dynamics have largely given up the requirement of direct reference to actual field situations. The object of a classical population dynamical model can be a fictitious population or a number of them, each assigned only a few attributes representing general properties such as birth and death rates, interaction coefficients, etc. – the attributes that are considered dynamically most relevant in a given context. The typical questions that these models are suitable to answer relate to the qualitative properties of the differential or difference equations that govern the dynamics of the model system; typical answers express criteria for the existence of equilibria, and specify their stability properties if they exist. Such dynamical models are called **strategic**, their aim being not to reproduce or to predict a concrete field situation, but to delimit the range of possible consequences for a certain set of simple assumptions, initial and boundary conditions – that is, for an ecological mechanism devoid of effects other than those defined in the model.

Strategic models are the means for obtaining a deep understanding of the relatively simple ecological mechanisms they are meant to represent – but these are hardly ever found in their pure form in nature. It is, therefore, very difficult to cite examples for simple, low-dimensional population dynamical models that have proved really helpful in reproducing or predicting actual field or experimental situations. For that end, it is almost always necessary (but, to be honest, not always sufficient) to consider many more biologically relevant details of the specific process to be simulated. The model thus obtained is a **tactical** representation of the concrete situation.

Apart from a very limited number of exceptions, classical theoretical population dynamics is purely strategic in the above sense. Different strategic models could produce almost any temporal coexistence pattern that appears on the field, but comparing these to actual field data, one could never be sure that the mechanism specified in the model is really an accurate representation of the one acting on the field. Completely different mechanisms might produce the same phenomenon – just think of the textbook example of the oscillating hare–lynx system, which can be explained very convincingly with at least three fundamentally different models. For example, the neutrally stable 'prototype' Lotka–Volterra model of predation (Section 3.4), a delayed density dependence model of the prey population with weak predator effect, or a driven system with one or more cyclically changing environmental factors are all capable of producing the same phenomenon of a regular out-of-phase oscillation of prey and predator. To decide – or at least to guess on an acceptable basis – which is the actual mechanism behind a certain temporal pattern, one has to compare the model parameters to those measured on the field. This is practicable only if the model parameters are relatively easily measured, which is unfortunately not the case for most strategic population dynamical representations.

The gap between the field-orientated practice of biosociology and the model-orientated theory of population dynamics has become discouragingly wide in this century, both disciplines having followed their own attitudes and methodologies for a long time, without much reference to each other's results. Now that the structural aspects of population and community dynamics on the one hand, and the dynamical constraints on community pattern formation on the other are clear to most ecologists, the need to bridge the gap seems to have become urgent for both sides. When considering spatial and temporal patterns of populations and communities, one can no longer be reliably interpreted without the other. This book is intended as a comprehensive survey of spatiotemporal population and community dynamical models representing the first operative steps towards a unified treatment of structure and dynamics in ecology.

Tamás Czárán
Budapest

Book structure

At the end of this section I shall provide a diagrammatic classification of spatiotemporal population dynamical models, based on theoretical considerations explained here and the corresponding model typology. The order and structure of presenting the models in the text corresponds to this classification in the first place.

Part One includes four different classical approaches, the common feature of which is that they always assume mass interaction within and among populations: no discrete objects like individuals are interpretable in such **mass-interaction models**. Two classes of them are **spatially explicit** in the sense that they represent topographic space either as continuous spatial coordinates (**reaction–diffusion** models, Chapter 2) or as discrete indices of locality (**patch-abundance** models, Chapter 3). The other two classes (**metapopulation** systems and **aggregated interaction** models, Chapter 4) are **spatially implicit**, because they do not address topographic space in any explicit way, but assumptions regarding a certain spatial structure of the biotic interactions are always hidden in their postulates.

Part Two includes population dynamical applications of spatial **object-interaction** systems, that is, **neighbourhood models**. These are all spatially explicit, but they differ from the classical models in that biotic interactions are assumed to take place among discrete objects (e.g., individuals) that are allocated to topographic space. Chapter 5 deals with **site-based** approaches (**cellular automata** and **interacting particle systems**). Chapter 6, the one including **individual-based** neighbourhood models (**tessellation** models and **distance** models), is the most diverse in relation to the number of different model types presented. The class of distance models incorporates three different approaches: **fixed radius neighbourhood** models, **zone of influence** models and **ecological field** models.

The order of presentation also roughly follows the historical order of the appearance of different model types. Along this 'historical gradient', a few canonical trends can be detected, like the approximate decrease in the spatiotemporal scale of the relevant dynamical processes, the increase in the number of biological facts considered in the models and the growing importance of numerical techniques in their analysis. Mass-interaction systems are mainly strategic, relatively large scale in both space and time, and analytical, whereas neighbourhood models are mostly tactic, spatiotemporally small scale and numerical. Of course, exceptions to these statements are easy to find in each model category, but this is the general trend.

Most models discussed in this book have been published elsewhere; the

Book structure xv

original sources are always cited. Of course the selection of models included is far from complete – I chose a few typical examples, but I had to omit many interesting systems, which could fill many times the number of pages of this volume. Some of the models – those in Sections 5.3.3, 5.3.6, 5.3.7 and a part of 6.2.2 – have not yet been published as research papers; these are either in press or are being prepared for publication.

Some of the formal derivations of certain models are boxed, in order to separate them from the main body of text; a few of the less trivial mathematical techniques used are described briefly in the appendices. However, readers uninterested in the exact mathematical structure of the models will be able to follow the main lines of argument without the boxes and the appendices. Each chapter ends with a brief summary, a synopsis of the most important results and conclusions in that chapter.

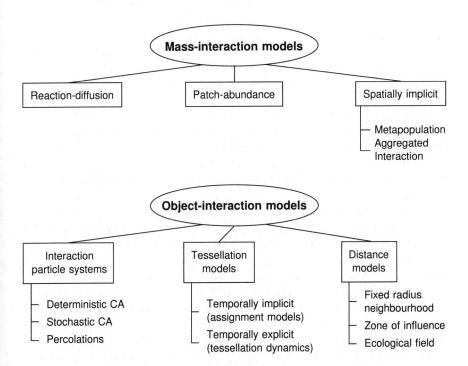

Acknowledgements

The idea to write this book had already been my secret plan for quite a long time before I received the honour of being invited to do so from Michael B. Usher, Principal Editor of the series, and Bob Carling, Senior Editor at Chapman and Hall. I am very grateful to them for initiating and taking care of the project. The happy coincidence of wish and possibility put me on the way of collecting material, and writing the book – both at a speed somewhat slower than I expected. Now, having finished the work, I deeply understand the message of the rule of thumb I heard from a mathematician friend: the first guess of the author concerning the time needed to produce a manuscript should be multiplied by a factor π to get a lower estimate for the actual time of delivery. I wish to thank the editors for their patience and kind encouragement throughout the work, and also for their help on many details.

Don DeAngelis has read the whole manuscript – the invaluable comments and suggestions he made were of great help to me in improving the text, sometimes by adding new perspectives or details, sometimes by making parts of the text more concise or clear. Suzanne Cordon has weeded out many little mistakes that I was not aware of; I wish to thank her for the thorough and accurate work she has done.

Many colleagues and friends have criticized and commented on the text, or on parts of it. I am especially grateful to Sanyi Bartha, with whom I started working on spatiotemporal ecological systems; I thank him also for the many exciting overnight chats about matters close to – or even very far from – the subject of this book. Many thanks are extended to Josef Hofbauer for his help with the material in Chapter 3, to Miklós Farkas for his helpful comments on Chapter 2, and to Ulf Dickman, Hans Metz, Rolf Hoekstra, Eva Jablonka, István Scheuring, János Podani and Beáta Oborny for many wise suggestions on various parts of the text.

In 1995, I had the privilege to spent 10 months in Collegium Budapest, Institute for Advanced Study, as a fellow of the Theoretical Biology Focus Group of the Collegium, in an exceptionally favourable environment for scientific work and interdisciplinary discussions. This was the most productive period in writing the book, partly due to the inspiring atmosphere of the Collegium, and partly because of its excellent library service run by Dorottya Bakó and Edina Földesi. Thanks for all these are due to the Collegium, especially to Lajos Vékás, the rector, and Eörs Szathmáry, the organizer of the Theoretical Biology Group.

Writing a book may be occasionally an exciting experience for the author, but it definitely requires a generous sacrifice from the family concerned. I thank my wife, Judit and my children, Dorottya, Domonkos and Boglárka, for their loving patience when missing me or seeing my back before the computer day in, day out. Thanks also to my mother for her moral and culinary support.

This project was supported by a number of research grants: OTKA T019524, T012793, T019364, T016096 and MKM 772, which I wish to acknowledge here.

1 Introduction

1.1 THE SIMPLIFYING ASSUMPTIONS OF CLASSICAL POPULATION DYNAMICS

The strategic models of classical, non-spatial population dynamics are based on three fundamental assumptions:

1. populations consist of large numbers of individuals (abundance assumption);
2. all individuals of the same population are identical in every dynamically relevant respect (uniformity assumption); and
3. the movement of the individuals is such that the population as a whole can be treated as an ergodic system (ergodicity assumption). (The ergodicity condition roughly refers to perfect spatial mixing in this context: each particle (molecule, individual) 'feels' the same environment around it, because the particles move fast and independently one from the other. Thus they collide (interact) with probabilities proportional to the products of their densities – or, in more chemical terms, their concentrations.)

Assumptions 1–3 are equivalent to the postulates of statistical mechanics and chemical kinetics – disciplines spectacularly successful in explaining many aspects of the macroscopic behaviour of gases and well-mixed liquid solutions, starting from the microscopic properties of the particles of which they consist. The 'mass action' type rules of chemical reaction kinetics are directly derived from the stoichiometry of chemical reactions and the models of statistical mechanics, the latter being built on the general assumptions 1–3, the laws of classical mechanics and statistical principles. Individual particles being indistinguishable, the fate of any single particle is irrelevant and impossible to follow.

Relying on the general assumptions in common with those of statistical mechanics and chemical kinetics, many methods from the powerful mathematical arsenal of physicochemical disciplines were naturally adapted to attack population dynamical problems. This led to a proliferation of phenomenological population models, the best-known classical examples of which are the Malthusian model of exponential population growth, the Pearl–Verhulst model of logistic growth, and the Lotka–Volterra models of population interactions. These, and a lot of other models, yielded deep theoretical insight into many aspects of population processes, which, beyond doubt, could not ever have been achieved without them.

2   Introduction

On the other hand, however, the lack of a broader interface between theoretical population dynamics and field ecology can be in a large part attributed to the fact that assumptions 1–3 do not apply to biological objects as well as they do to molecule populations. The uniformity assumption is violated in every actual field case, almost without exception, in population biology: individual variation within biological populations is much higher than that of the particles in a usual physical or chemical system. One can say that, in fact, every individual of a community should be regarded as unique for its combination of dynamically relevant genetical, physiological and environmental properties, even when comparing individuals of the same species (Huston *et al.*, 1988; DeAngelis and Gross, 1992). Ecological objects also violate the abundance and the ergodicity criteria: the number of individuals is fewer by tens of orders of magnitude, even in a very dense community, than the number of particles in a test tube, and their motion is usually very far from ergodic on all spatiotemporal scales. The ergodicity condition is spectacularly violated by plants and benthic animals, which do not move at all through most of their lifetime: they can interact with close neighbours only, and practically they do not feel the presence of individuals outside a limited spatial interval around them. But even populations of animals capable of very fast movement do not mix perfectly: individuals might stick to home ranges, mates or resource patches to some extent; thus they experience more or less neighbourhood effect as well, albeit possibly on a larger spatial scale and in a fuzzier manner than sessile organisms do. Many of the abundant cases of even qualitative misfits between the predictions of classical population dynamical models and actual field data can be attributed to the violation of the ergodicity assumption. Of course, neither statistical mechanics nor chemical kinetics suffer so much from the discrepancy between theoretical expectations and empirical observations because their usual objects approximately satisfy assumptions 1–3: they are really huge sets of identical particles, which can be safely assumed to mix well and move fast enough to 'average out', on the macroscopic spatiotemporal scale, the effects of the local context on the microscopic level.

Assumptions 1–3, which proved to be sufficiently realistic in models of simple physicochemical systems, are, in fact, imposed by a methodological constraint on population dynamics. Mass-action type population interaction rules are necessary to postulate, because these allow for mathematical representations simple enough for analytical treatment. Even then, the number of interacting populations, that is, the dimensionality of the system, remains a critical methodological problem of classical population dynamics. Analytical tractability requires this number to be either small (not larger than three or four in most cases) or infinitely large. For the number of populations to be acceptably low, many individuals are to be lumped together, based on their similarity regarding a dynamically relevant characteristic, or a combination of such characteristics. What individual characteristics make the basis of the grouping is determined by the question the model is expected to answer. It can be the species identity, age, developmental stage, size or the spatial position of the individuals, for example, among many others. At one extreme, all individuals can be assumed identical, which is the case in the single-species, unstructured models of Malthusian or logistic population growth, for example. Following the conceptual framework and the terminology of Metz and Diekman (1986) and Caswell and John (1992), we say that all individuals are in the same **individual state** or i-state in these models.

A Lotka–Volterra type competition model, for example, represents interactions among populations that differ in species identity. This is the only relevant i-state variable (descriptor) of the model, and it takes s different values (s being the number of species considered), so that the **i-state space** of such models is one dimensional and discrete. All individuals belonging to the same species are assumed identical, forming a homogeneous mass – in fact, the individuals as discrete entities are not considered in such classical models at all. The basic objects are the populations themselves, labelled by their species names as the corresponding i-state values. It is meaningless to ask anything about certain individuals; the relevant dynamical questions regard the **abundances** of the competing species. All mechanisms of population growth and competitive interaction are formulated in terms of species abundances, and they affect species abundances directly. This situation is the analogue of a well-mixed chemical system in a fluid medium, in which s different compounds interact (react) with each other. The abundance distribution on the i-states is called the **population state** or p-state, which is, in the case of a Lotka–Volterra competition model, represented by the s-dimensional vector **X** of species abundances. Note that the dimensionality of the p-state space (in this example, s) is equal to the number of possible i-states considered, and **not** the dimensionality of the i-state space (which is one in the example).

The dynamical rules of the Lotka–Volterra model can be formulated on the p-state level, as a system of ordinary differential equations, of which the population densities x_i are the state variables:

$$\frac{dx_i}{dt} = r_i x_i - \sum_{j=1}^{s} \alpha_{ij} \cdot x_j x_i \quad (i = 1, \ldots, s)$$

The species interact through the 'mass action' type rules of classical population dynamics, which are analogous to those of chemical kinetics: the intensity of the interaction between populations i and j is proportional to the $x_i x_j$ product of their densities (concentrations). The parameters of the model are p-state parameters, which characterize the speed of population growth (r_i) for species i when it is not constrained by competition, and the competitive effect (α_{ij}) that a unit density of population i suffers in the presence of a unit density of population j. The relevant questions these models can address are the stability properties of the p-state, that is, the stationarity conditions of the **abundance distribution** on the i-states (the reader will find some results for the Lotka–Volterra system in Sections 3.3 and 4.3). This is why Caswell and John (1992) and Maley and Caswell (1993) call such systems ***i*-state distribution models**, suggesting this term as a more precise substitute for the 'state variable model' concept of Huston *et al.* (1988). Based on a somewhat less abstract property that is common in all i-state distribution models, namely that they define the dynamical relations of the populations in terms of mass interaction, I shall use the more intuitive term '**mass-interaction model**' as a synonym for the 'i-state distribution model'.

If more than one i-state descriptors make the difference among individuals (for example, the identity and larval stage of both species are dynamically relevant), then a particular i-state can be thought of as a point in the i-state space (which is then two dimensional (2D) and discrete in both dimensions), spanned by the i-state descriptors as axes. Therefore the p-state representation is not a vector distribution, but a higher-dimensional distribution of abundances (a 2D matrix distribution in the example). This makes a difference in analytical tractability, as the number of possible i-states, and thus the number of simultaneous equations, might be large but the resulting model is still a system of ordinary differential equations (ODEs). If, however, at least one of the i-state descriptors take a continuum of values (e.g., age in continuous time), then the i-space is also continuous in the corresponding dimension, and the i-state distribution (mass-interaction) problem becomes infinite dimensional. The appropriate mathematical tool to attack this is a system of partial differential equations (PDEs).

1.2 SPATIAL EXTENSIONS OF THE CLASSICAL APPROACH

The consideration of spatial aspects in population dynamics allows for the relaxation of at least one of the biologically unrealistic 1–3 postulates: spatial population models never satisfy the ergodicity assumption on all scales. In the most general terms, this means that overall spatial mixing, thus the spatially homogeneous growth and interaction of the populations, is replaced by a **spatially constrained** mechanism. How spatial constraints are imposed on the populations depends on the actual type of the model. A straightforward way of doing this is to extend a non-spatial mass-interaction model – for example, a Lotka–Volterra type model – so that the i-states of the subpopulations include spatial position as well. If space is continuous, then the i-state descriptors of spatial position are Cartesian coordinates, that is, continuous variables; if space is discrete, the corresponding i-states are discrete identifiers of locality (index names of islands, habitat islands). In the former case, the model is a **reaction–diffusion system**, represented by a system of PDEs. In the latter, we have a **patch-abundance model**, represented by a system of ODEs, one equation for each i-state. Reaction–diffusion and patch-abundance models are both mass-interaction systems, since it is still the abundance (density) of the populations that is directly affected by the dynamics – but the local, not the overall abundance, unlike in non-spatial models. In fact the interactions of individuals **within** the localities are ergodic in these systems as well, only the large-scale movements **between** the localities are spatially constrained.

If space is not in any explicit way included among the relevant i-states, but the model implicitly assumes a spatial structure of the habitat, affecting the abundance dynamics of the resident populations, we have a spatially implicit mass-interaction model. **Metapopulation** (that is, patch occupancy) and **aggregated interaction** (mainly host–parasitoid) models are the basic categories of this class. Metapopulation models fit into the non-spatial mass-interaction framework, with the conceptual modification that it is the habitat patch that takes the role of the individual, and the whole set of patches is analogous to the population. Thus, the fate of a single habitat patch is irrelevant and impossible to trace in a metapopulation model. The analogy can be so close that metapopulation models are often formally equivalent with certain non-spatial population models (cf. Section 4.2.1). Examples for i-state descriptors (variables) in a metapopulation model are resident species combination and patch age, to take just the most obvious ones; it is the abundance distribution of the patches on the i-state space that is directly modelled.

Aggregated interaction models – the other type of spatially implicit mass-interaction systems – represent a very different approach. These incorporate the effects of spatial habitat structure by manipulating the interaction terms of originally non-spatial models, so that the dynamical consequences of the aggregation of interactions can be assessed without any explicit reference to space. Kareiva (1990) called these models 'pseudospatial'.

1.3 THE OBJECT-BASED APPROACH: NEIGHBOURHOOD MODELLING

A common feature of all mass-interaction models – spatial and non-spatial alike – is that they conform to the general assumptions of the high abundance and the uniformity of individuals within any possible i-state. It is quite a recent development of theoretical ecology that these assumptions are no longer unavoidable. There is a fast developing class of models directly based on the fact that any two individuals do differ in at least one aspect from each other: even if they are, unrealistically, postulated to be identical – for example in their genetic constitution, physiology or environment, the similarity of their i-states is always constrained spatially and temporally. To be short, two individuals cannot be present at the same place simultaneously.

Given the inherently local manner of the interactions among individuals, the positional aspect of their i-states might be of much dynamical relevance, especially on smaller spatial scales. This is an aspect of i-state that is necessarily omitted in mass-interaction models, which always assign a large number of individuals to a single spatial point – this is most obviously the case in non-spatial systems, but the same applies to patch models and diffusion models as well. If spatial position is dynamically important even on a scale close to the size of a single organism, the individuals cannot be legitimately lumped together into large, internally homogeneous groups, and no such groups can be regarded as the units of interaction. On the contrary, the subdivision of the population must go down close to the level of single individuals in order to find the ultimate, internally homogeneous unit of interaction. Then each such interacting unit must be characterized by its own i-state, differing from those of all the others. The most extreme case is when the interacting unit is the individual itself, so that the number of i-states equals the number of individuals. But the interacting unit needs not be the solitary individual at all. It can be, for example, a small group of individuals belonging to the same family, a small colony occupying the same habitat patch, or a tussock of grass, the ramets of which belong to the same genet (clone). i-State distribution approaches are obviously inadequate to represent the dynamics in such situations, because the criteria for the application of the statistical principles underlying the mass-interaction formalism are seriously violated, e.g., by the uniqueness of the interacting units. At this level of dynamical resolution, the rules of interaction must be defined among the unique objects themselves, not among the densities of internally uniform groups of individuals. The dynamical rules act on the i-state level, altering the i-states of the interacting objects directly, unlike in mass-interaction models, where the dynamical rules act on the **abundances** of the i-state classes, that is, on the p-states.

Caswell and John (1992) call the i-state-based interaction rules **constraint functions**, and the population-level representation of i-states, in a certain environmental setting, an ***i*-state configuration** (see also Maley and Caswell (1993)). The i-state configuration of a given system is a list of the individual objects it includes, together with their i-states. A simple example for an i-state configuration is a map showing the spatial positions of all individuals, each assigned all the i-state attributes (e.g., species identity and/or age, size etc.) that are considered as dynamically relevant in the model of interest.

An i-state configuration type p-state is obviously more informative than is an i-state distribution type p-state for the same set of individuals, since the number of i-states considered is equal – or at least it is close – to the number of individuals in the former, whereas it is much less in the latter case. Therefore it is always possible to create a number of i-state distributions from an i-state configuration by simply omitting some i-state descriptors and lumping individuals together differing in those only, but the reverse cannot be done without relying on additional information. In the above example, omitting the spatial coordinates of the individuals would yield an i-state distribution, with individuals of the same species (age, size, etc.) belonging to the same group.

The constraint function is the set of rules defining the way i-state configurations are transformed in time. It can be a function (either deterministic or stochastic) of any one, or any combination, of the i-state descriptors. As stated before, the constraint function acts on the level of the individual objects, by altering i-states directly. Such dynamical systems are called ***i*-state configuration models** by Caswell and John (1992), as an extension of the less general concept of **individual-based models** (Huston *et al.*, 1988). I propose the more intuitive but equally precise '**object-interaction model**' synonym for the term 'i-state configuration model'.

If the argument of the constraint function includes spatial position, the object-interaction model is explicitly spatial; if spatial proximity is one of the factors determining the strength of the interaction between the objects, one has a **neighbourhood model** in hand. Neighbourhood models are a subset of the spatial subset of object-interaction models: an object-interaction model need not be spatial, and a spatial object-interaction model need not be a neighbourhood model. For example, morphological constitution, physiological status, age, developmental stage, motivational state, or any other i-state descriptor might determine interaction among individuals, without any direct reference to spatial position. Even if there is spatial reference in the transformation rules, it might not be limited to a certain neighbourhood around the individual objects. Thus, there is a wealth of potentially different object-interaction systems. It is a sign of the importance of spatial proximity in ecological interactions that by far the most object-interaction approaches in ecology belong to the specific class of neighbourhood models.

Neighbourhood models developed so far can be naturally and conveniently classified into four major groups, according to their spatiotemporal set-up and regarding the constraint functions:

1. cellular automata (CA);
2. interacting particle systems (IPSs);
3. tessellation models (TMs); and
4. distance models (DMs).

A rough characterization of these model types can be based on the way they define neighbourhoods and the nature of their interaction rules. Figure 1.1 provides examples for some of the most common neighbourhood definitions (cf. Czárán and Bartha, 1992).

From the theoretical viewpoint, interacting particle systems and cellular automata differ from tessellation models and distance models in at least two important respects: (i) IPS and CA models are spatially discrete, whereas tessellations and distance models are spatially continuous; (ii) in fact the real interacting object of IPS and CA models is the discrete spatial unit (site) within the grid of such units, whereas the interacting unit is the individual itself in the other two types of neighbourhood models.

The second difference is a fundamental one: in IPSs and CA it is always a unit of space, that is, the **site** (or cell) to which different i-states are assigned, and it is only in specific cases that this i-state is an attribute of a

8 Introduction

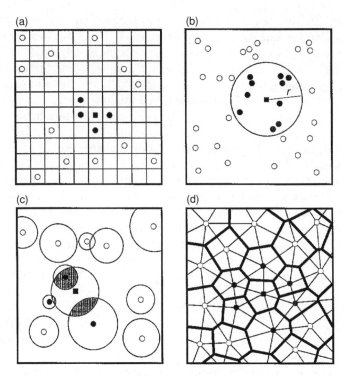

Figure 1.1 Neighbourhood definitions: (a) cellular automata and interacting particle systems; (b) fixed radius neighbourhood models; (c) zone of influence models; and (d) tessellation models.

single individual occupying the site. It is the fate of the site, not of the individual, that is followed in IPS and CA models; site states can be – and in most models they are – different from the state of a single individual. Thus, in the strict sense, the name 'interacting **particle** systems' is somewhat misleading: the physical applications of IPS models are **not** particle based, and the ecological applications are **not** individual based. State descriptions and state transitions of IPSs and CA apply to localities directly, therefore the i-state variable can be different from anything applicable to a single individual. It can be, for example, the species composition occupying the site if the spatial resolution of the model requires that, as in CA implementations of metapopulation models (Caswell and Etter, 1993). In some other systems the relevant state of a site is the (small) number of individuals it actually contains (e.g., Weiner and Conte, 1981; Czárán, 1989; Palmer, 1992). Even if the site-to-individual correspondence is unambiguous at a certain point of time, it need not stay the same by the next generation: a mobile organism might move from one site to another (as in the model of De Roos *et al.* (1991) for example), whereby it alters the i-state of two sites simultaneously: the one it leaves and the one it arrives at. It is also possible that an individual occupies more than one site of the grid at the same time, as happens to be the case in models of clonal plant growth (e.g., Ford, 1987; Callaghan *et al.*, 1990; Oborny, 1994a). The mapping of individuals onto the lattice may be variable both in space and in time.

Unlike in IPS and CA models, the basic interacting object of a TM or a DM is **always** the solitary individual, the i-state of which always includes its spatial position. This difference serves as the basis for classifying interacting particle systems and cellular automata as **site-based** neighbourhood models, while tessellation models and distance models are discussed under the heading **individual-based** neighbourhood models in the sequel. There is also a difference between these two types in the mathematical representation of space: site-based models are spatially discrete, whereas the overwhelming majority of individual-based models are spatially continuous. Tessellation models allocate a part of the habitat space to every individual, whose dynamical performance is a direct function of the area and the shape of the captured habitat space segment. The allocation of space can be implemented in many different ways, depending on the spatial, temporal and physiological constraints of the organisms considered. In distance models, the pairwise interaction between any two individuals is a direct function of their distance in space – but the actual interaction functions can be of very different forms.

Neighbourhood models got established as standard theoretical tools in almost all branches of natural sciences in the past decade, from theoretical physics through geography to many branches of biology. There are also strong signs of their penetration into economics and the humanities – sociology, for example – wherever the spatial interactions of unique entities can be an issue of interest. Besides their popularity as model systems in many branches of science, interacting particle systems, cellular automata and tessellations are eminent fields of research for a new, thoroughly computerized mathematical discipline called experimental mathematics, which combines analytical and numerical techniques in the study of the formal structure of these (and other) models.

Existing neighbourhood models are very diverse both conceptually and structurally. There is one methodological point in which they are similar, however, and that methodological similarity reflects a deep structural homology. Principally owing to the fact that the fate of each individual object is kept track of through time in neighbourhood models, there is hardly any chance to analyse them by pure analytical techniques – the most adequate (and almost always the only possible) means of discovering the dynamical behaviour of a neighbourhood system is computer simulation. Analytical approximations are feasible in specific cases, but these usually presume the omission of very important spatial aspects (e.g., the variation in neighbourhood composition). The consequence of the loss of analytical tractability is an inevitable loss in the generality of results and thus of the theoretical insight obtainable (Lomnicki, 1992; Metz and DeRoos, 1992; Murdoch *et al.*,

10 Introduction

1992), but many of the recent theoretical problems of population and community dynamics cannot even be formulated, let alone solved, without an object-interaction modelling framework (cf. Judson, 1994; Wolff, 1994; Conroy *et al.*, 1995; Dunning *et al.*, 1995; Holt *et al.*, 1995; Turner *et al.*, 1995).

1.4 THE ROLE OF SPATIOTEMPORAL MODELS IN ECOLOGY

Due to considering spatial population structure on a wide scale of different approaches, spatiotemporal modelling is perhaps the most adequate means of bridging the methodological gap between theoretical and field ecology. Reaction–diffusion and patch-abundance models – being direct extensions of the classical non-spatial systems – are ideally suited for assessing the genuine effects of spatial structure on population dynamics by comparing the predictions of the spatiotemporal systems to those of their non-spatial counterparts, all other things kept equal as much as possible. The reaction–diffusion and the patch-abundance analogues of certain non-spatial systems might qualitatively differ in their dynamical behaviour both from the non-spatial version and from each other; the explicit inclusion of the spatial aspect often leads to more realistic coexistence patterns. Metapopulation and aggregated interaction models (that is, spatially implicit systems) are routinely tested against field data, but at the same time many of them provide valuable theoretical insights as well. Almost the same applies to site-based neighbourhood models, the important difference being that while a spatially implicit model can often be relevant theoretically and practically alike, almost any concrete cellular automaton or interacting particle system model may be quite securely categorized as either theoretically or practically motivated (that is, strategic or tactical). It is the class of individual-based neighbourhood models that is closest to field practice both in its characteristic spatiotemporal scale and in the potential realism of the biological mechanisms that a distance model or a tessellation model can incorporate. The majority of existing individual-based models are tactical for the time being, albeit the possibility of applying them to more general theoretical problems is open. With distance models and tessellation models, population dynamics has reached the dynamical resolution level where computer-simulated processes are directly comparable to actual field processes on the smallest relevant spatiotemporal scale.

Part One
Spatial Mass-interaction Models

2 Reaction–diffusion models of population growth and dispersion

2.1 INTRODUCTION

During almost a century of development initiated by the pioneering work of Pearson and Blakeman (1906), biodiffusion has become one of the most diverse and elaborate topics of biomathematics. The present prosperity of diffusion modelling (following the loose terminology of the literature, the term 'diffusion model' designates all kinds of population dynamical models with a diffusion component included, even if they address other component processes (advection, population growth, interactions) as well) in biology is indicated by thousands of research papers and excellent monographs and reviews (Okubo, 1980; Berg, 1983; Britton, 1986; Murray, 1989) recently published on a wealth of different aspects of biodiffusion. In theoretical population dynamics specifically, the persistent interest in reaction–diffusion systems is not simply a consequence of their historical priority within the widening spectrum of spatiotemporal approaches. It is maintained by the unceasing yield of interesting and often counter-intuitive theoretical results that diffusion modelling still produces in theoretical ecology. In addition, specific reaction–diffusion models have proven to be sufficiently realistic to explain actual abundance and distribution patterns in terrestrial and aquatic systems, from the spatiotemporal distribution of planktonic organisms (Kierstead and Slobodkin, 1953; Possingham and Roughgarden, 1990) through the spread of certain epidemics (Diekmann, 1978; Murray *et al.*, 1986; Yachi *et al.*, 1989; Pech and McIlroy, 1990) to that of populations of higher organisms (Ludwig *et al.*, 1979; Fleming *et al.*, 1982; Kareiva, 1983; Kareiva and Shigesada, 1983; Okubo *et al.*, 1989; Andow *et al.*, 1990; Lawton and Godfray, 1990), even though most diffusion models are motivated by theoretical problems in the first place.

Following the general verbal definition given by Okubo (1980), **diffusion** is regarded as a regular dispersion movement of groups of particles (in our case, populations) arising from the irregular, e.g., Brownian-like motion of the particles (individuals) themselves. If the movement of the particles is directionally affected by an external or internal force, for example by a flow of the medium (e.g., air or water current) or the motivation to move towards an attractive point in space (such as a rich source of food or a potential mate), the resulting directional component of the dispersion movement of the group is called **advection**. Besides dispersion movement, particles might also appear and disappear through demographic events (births and deaths) or as a result of interaction with other particles. Demographic and interaction events are collectively called **reactions**, using the terminology of chemistry.

It might seem self-evident from these definitions that diffusion models can only be applied to populations of mobile organisms, since the movement of individuals is an essential part of the postulates. For this reason, most strategic models of reaction–diffusion dynamics are motivated by zoological problems, and most known terrestrial applications are zoological. Two of the rare exceptions are Cain (1990) and Allen *et al.* (1991). It is important to see, however, that this is by no means a rule: the only criterion required for diffusion models to be applicable is that the individuals be mobile at some stage of their life cycle. Even populations of sessile vascular plant or benthic animal species may be seen as dispersing by diffusion, if the irregular movement of a single individual is substituted by the irregular path of the successive generations of mobile propagules. For vascular plants it is the seeds, stolons, rhizomes, etc. that disperse along more or less irregular paths, and the resulting movement may be very much like random walk. The same applies to sessile benthic animals, whose planktonic larvae might themselves move in a diffusive–advective fashion in the fluid medium, before they settle down to reproduce. This kind of diffusion is perceivable on a larger spatiotemporal scale than that of a mobile animal population, since each single step of movement may be quite long both in time and in space, measured sometimes in hundreds or even thousands of kilometres, and lasting for a full generation time.

Diffusion models relax one of the most unrealistic basic assumptions, inherent, without exception, in classical (e.g., Lotka–Volterra type) models of population dynamics: that of overall spatial homogeneity regarding all relevant variables and constants, and the perfect mixing of the population on all spatial scales. Dynamical events are spatiotemporally determined, that is, population abundances and demographic parameters are defined as direct functions of spatial and temporal coordinates. Potentially the same applies to any external (milieu) variables, which enables the dynamical effects of environmental inhomogeneities or heterogeneities to be studied. Demographic and migratory processes are postulated to be of limited spatial range and population interactions to be local, thus excluding the assumption of direct spatially long-range dynamical effects that are in fact quite unusual between biological entities, but far too common in models devoted to represent them. The price to be paid for these advantages is methodological: it is at times quite laborious, if at all possible, to handle by purely analytical means the partial differential equation formalism some problems may require, even for those fairly confident in handling complicated mathematical objects.

Motivated by the theoretical advantages as compared to classical approaches, diffusion dynamics has attracted intense professional interest, resulting in so rich an output of review and textbook literature on the subject thus far that it would be both unproductive and pretentious to try re-reviewing them in a single chapter. What follows is therefore by no means an attempt to summarize all the results of diffusion modelling, but a selection of some interesting models, with the postulates, important background assumptions and typical results of some basic types explicated as simply as possible. My bias is definitely towards the logic of formulating diffusion models, and not the technical details of analysing them. With a reader possessing experience in secondary school level calculus and probability theory in mind, I intend to keep formal details at a level necessary to get an insight into the way diffusion models are set up and handled, and to understand the argumentation of this and the remaining chapters. Full derivations are confined to the routes between postulates and model equations, in order to clarify what the models are able to explain and where the limits of their applicability are. For more details on the formal treatment (solutions, stability analyses) of partial differential equations and diffusion models specifically, the reader can turn to any of the mathematical textbooks devoted to the topic (e.g., Tychonov and Samarski, 1964, 1967).

2.2 RANDOM WALK APPROXIMATIONS TO DIFFUSION

Following the methodological instruction implicit in the definition of diffusion, the collective behaviour of a dispersing population can be approximated through the assumption that each individual moves randomly. The best-known physical example of random movement is realized by microscopic solid particles dispersing in a fluid medium – a process called Brownian motion. The simplest possible model representing Brownian motion is that of **random walk** in one dimension, the subject of this section. Considering a population of individuals moving along irregular paths without any direction or spatial point preferred, the analogy with this physical 'prototype' realization of random movement becomes apparent (Berg, 1983). When we try to utilize this analogy in setting up population dynamical models, it is not to be forgotten, however, that by so doing we attribute a very limited repertoire of physiological and behavioural capabilities to the individuals of the hypothetical population, which should and can be improved (Marsh and Jones, 1988; Turchin, 1988) to obtain more generally interpretable models.

2.2.1 Random walk in discrete time and space

Imagine an infinite one-dimensional array consisting of discrete spatial units of length λ each, along which a hypothetical population of identical individuals exist. During observation, no demographic events (births, deaths,

16 Reaction–diffusion models

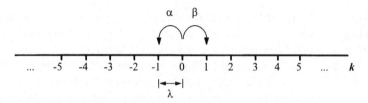

Figure 2.1 Isotropic random walk in one dimension.

immigration, emigration) occur. At time t_0, the population is released from point 0, the origin. We suppose that all individuals step into one of the spatial units (cells) next to their previous position in a time unit of duration τ. The probability of stepping to any one direction (left or right) is $\alpha = \beta = 0.5$, that is, the movement of individuals is isotropic, and in each time unit they step exactly once (Figure 2.1). Now we ask how the population is distributed along the spatial axis after n time units have elapsed? To obtain the distribution in analytical form, we have to determine the probability $P(k, n)$ that an individual, starting from the origin and taking n steps, gets into a cell k spatial units away from the origin.

We derive the probability of reaching the kth spatial unit to the right of the origin (negative k values represent a net movement to the left from the origin). For this, the individual needs to realize r steps right and $(n - r)$ left, so that

$$k = r - (n - r) = 2r - n. \tag{2.1}$$

Rearranging,

$$r = (k + n)/2. \tag{2.2}$$

Note that for r to be an integer, $(k + n)$ must be even, that is, there is a non-zero probability of reaching a spatial unit only if k takes its value from

$$\{-n, -(n-2), \ldots, (n-2), n\}. \tag{2.3}$$

Equation (2.3) implies that the parity of all possible k values is the same as that of n, which follows from the assumption of each individual stepping exactly once in each time unit. The probability we look for is the number of possible paths that consist of exactly r rightward movements, divided by the total number of possible paths, that is, the fraction of favourable series of steps within the complete set of possible series. The number of possible paths is easy to determine: each of the n steps can be one of two possibilities: rightward or leftward, that is, the number of possible paths is 2^n. The number of different paths ending k steps to the right of the origin may be determined by combinatorial reasoning as

$$C_r^n = \frac{n!}{r!(n-r)!}, \qquad (2.4)$$

the number of combinations for r rightward steps within a total of n steps.

Denoting by $P(k, n)$ the probability that an individual steps exactly r times to the right, $(n - r)$ times to the left, and thus gets $k = 2r - n$ spatial units away to the right of the origin by taking n steps, we have

$$P(k, n) = \frac{C_r^n}{2^n} = \frac{n!}{(n-r)!r!} \cdot \left(\frac{1}{2}\right)^n = \binom{n}{r} \cdot \left(\frac{1}{2}\right)^n, \qquad (2.5)$$

with

$$r = (n+k)/2 \quad \text{and} \quad k \in \{-n, -(n-2), \ldots, (n-2), n\}. \qquad (2.6)$$

$P(k, n)$ can be calculated from the binomial distribution.

$P(k, n)$ is a probability measure, but supposing a great number of identical individuals follow the same rule of random walk, each probability for the different possible k values between $-n$ and n will be realized by a fraction $P(k, n)$ of the whole population. Thus the population spreads along the spatial axis according to the binomial distribution with n increasing, as shown on Figure 2.2. The variance of the binomial distribution is known to be proportional to n (for this special case of isotropic random walk, it is in fact equal to n), implying that the spread of the population, as expressed by the variance of its spatial distribution, is a linear function of time.

2.2.2 Random walk in continuous time and space

This simple model of random walk can be extended so that, instead of the discrete spatial and temporal variables k and n, the resulting equation is continuous in the spatial and temporal coordinates x and t. For this end, the original discrete habitat array should be replaced by a continuous one-dimensional space, and the rules of individual movements should be somewhat modified to obtain the probability density function $\hat{p}(x, t)$ of the position of the individual at time t.

We assume that an individual released in the interval $[0 \pm l)$ centred on the origin, stepping n times altogether, r times to the right, gets into the interval $[x \pm l)$, where

$$x = k\lambda = (2r - n)\lambda. \qquad (2.7)$$

This rule comes from supposing that the exact position of the individual within the $[0 \pm \lambda)$ interval is not known (it may be anywhere within the interval with an evenly distributed probability density), and each step is of

18   Reaction–diffusion models

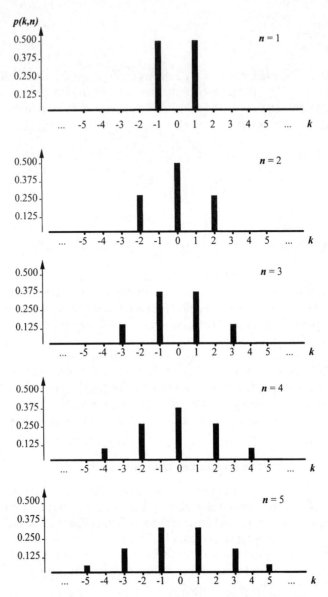

Figure 2.2 The (binomial) probability distributions of the spatial position of a particle randomly walking in one dimension; the first five time units.

length λ, directed left or right at random with $\alpha = \beta = 0.5$, as in the discrete case, as demonstrated in Figure 2.3. The rule is equivalent to that leading to (2.5), except that now we use continuous space as the 'habitat' of the hypothetical population. Rearranging (2.7),

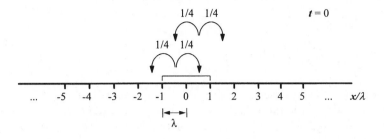

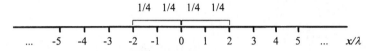

Figure 2.3 Isotropic random walk in continuous space: the spread of positional probability density in one dimension.

$$r = n/2 + x/(2\lambda). \tag{2.8}$$

n time units of duration τ add up to give

$$t = n\tau, \tag{2.9}$$

that is,

$$n = t/\tau. \tag{2.10}$$

Substituting (2.8) and (2.10) into (2.5) yields

$$P(x, t) = \frac{(t/\tau)!}{\left(\frac{t/\tau + x/\lambda}{2}\right)! \cdot \left(\frac{t/\tau - x/\lambda}{2}\right)!} \cdot \left(\frac{1}{2}\right)^n. \tag{2.11}$$

Now we suppose that $\tau \ll t$, that is, $n = t/\tau$, the number of steps taken is very large. Then the factorials in (2.11) will be large enough to be precisely approximated by the Stirling formula (e.g., Feller, 1968) in the form

$$n! = (n/e)^n \cdot \sqrt{2\pi n}. \tag{2.12}$$

Applying (2.12) to all the factorials in (2.11), and working through the necessary algebra results in

$$P(x,t) = \frac{1}{\sqrt{2\pi}} \cdot 2\left(\frac{t\tau}{t^2 - x^2(\tau^2/\lambda^2)}\right)^{-\frac{1}{2}} \cdot \exp\left(-\frac{x^2\tau}{2\lambda^2 t}\right). \qquad (2.13)$$

Equation (2.13) gives the probability that a single individual released at time $t_0 = 0$ from within the interval $[0 \pm \lambda)$ will get into the interval $[x \pm \lambda)$ in time t, or – supposing a huge number of individuals follow the same pattern of movement – $P(x, t)$ is the fraction of the population arriving at $[x \pm \lambda)$ by time t. To obtain $P(x, t)$ normalized for a unit of habitat length, that is, as a **probability density** measure, we divide it by 2λ, the length of the interval $[x \pm \lambda)$:

$$\hat{P}(x,t) = \frac{P(x,t)}{2\lambda} = \frac{1}{\sqrt{2\pi}} \cdot \left(\frac{t(\tau/\lambda^2)}{t^2 - x^2(\tau^2/\lambda^2)}\right)^{-\frac{1}{2}} \cdot \exp\left(-\frac{x^2\tau}{2\lambda^2 t}\right) \qquad (2.14)$$

Now assume that λ and τ decrease through all limits on the condition that λ^2/τ approaches a finite positive value $2D$ at the same time. (Note that having very small τ values is consistent with the above assumption of t/τ being large.) Taking the limits $\lambda \to 0$ and $\tau \to 0$, $\tau^2/\lambda^2 = \tau/(2D)$ vanishes in the denominator of the first bracketed term in the right-hand side of (2.14), so that we get

$$\lim_{\substack{\lambda,\tau\to 0 \\ \lambda^2/\tau \to 2D}} \hat{P}(x,t) = \hat{p}(x,t) = \frac{1}{\sqrt{4\pi Dt}} \cdot \exp\left(-\frac{x^2}{4Dt}\right). \qquad (2.15)$$

By comparing (2.15) to

$$\Phi_{m,\sigma}(x) = \frac{1}{\sqrt{2\pi\sigma^2}} \cdot \exp\left(-\frac{(x-m)^2}{2\sigma^2}\right), \qquad (2.16)$$

the probability density function (PDF) of the Gaussian (normal) distribution, $\hat{p}(x, t)$ turns out to be identical to $\Phi_{m,\sigma}(x)$ of $m = 0$ mean and

$$\sigma^2 = 2Dt \qquad (2.17)$$

variance. The positive constant D is called the diffusion coefficient or the diffusivity of the hypothetical population, characteristic of the speed of its spatial spreading. Note that D is **not** the speed of walk, which can be assigned to individuals, not to the population as a whole. The speed of walk (v) is determined by $v = \lambda/\tau$, which approaches positive infinity in the limit as $\lambda \to 0$ and $\tau \to 0$, since we assumed that it is λ^2/τ that remains finite. This peculiar feature of the model is also reflected in the fact that the Gaussian distribution (the limiting distribution of the binomial as n becomes very large) has positive values all along the $(-\infty, \infty)$ continuum, no matter how small its variance is. Thus, for any positive t, $\hat{p}(x, t)$ is positive (there is a non-zero probability of reaching x) even if x is infinitely far from the origin. Since we assumed the population to be released in $x = 0$, this is only possible if the speed of walk is infinite. Of course such an assumption is hardly justifiable on a biological basis – infinite speeds are rarely attained by animals, let alone plants. The only justification is that the resulting model gives a good approximation for the spreading process of populations of randomly moving objects if t is not very small (Okubo, 1980), and it is much more tractable than any discrete models of random walk.

The area below the graph of (2.15) between the spatial points a and b is given by

$$Q(\hat{p}, a, b) = \int_a^b \hat{p}(x, t) dx, \qquad (2.18)$$

which is the probability that an individual starting from $x = 0$ at $t = 0$ will be within the interval $[a, b)$ by time t, or – *mutatis mutandis* – the fraction of a large population released at $x = 0$ arriving into this interval by time t (Figure 2.4). The variance of the spatial distribution is again a linear function of time as can be seen from (2.17), much as in the discrete (binomial) model (Figure 2.5).

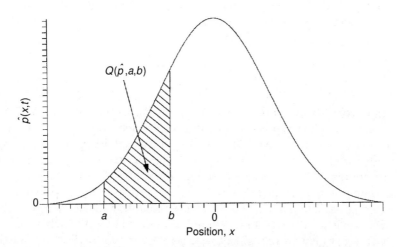

Figure 2.4 The probability density function (PDF) of the position of a randomly walking particle; $Q(\hat{p}, a, b) = \int_a^b \hat{p}(x, t) dx$ is the probability that the particle starting from $x = 0$ will be within the $[a, b)$ interval by time t.

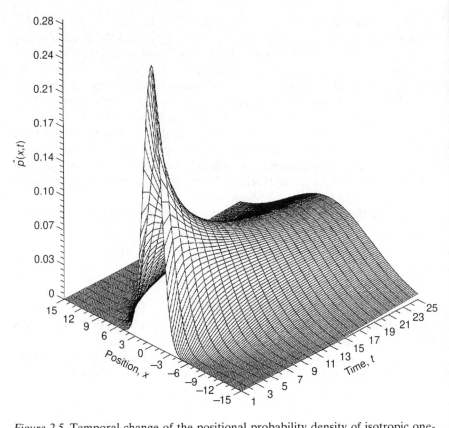

Figure 2.5 Temporal change of the positional probability density of isotropic one-dimensional random walk in continuous space and time.

2.3 THE DIFFUSION EQUATION

Starting from a hypothetical microscopic mechanism of random walk, we obtained an explicit expression (2.15) directly applicable to the determination, at any instant t, of the PDF of getting to any spatial point x for a single individual starting at $x_0 = 0$. According to the alternative population-based interpretation of (2.15), $\hat{p}(x, t)$ describes the spatial distribution of a large spreading population for any point of time, provided the population was released at a single point $x_0 = 0$. This latter condition means that the initial distribution in $t_0 = 0$ is the Dirac-δ function centred at $x = 0$, which we denote by $\delta(0)$. ($\delta(0)$ may be thought of here as the limiting PDF of a Gaussian distribution with zero mean and zero variance.) To get a more general description not confined to the specific (and rather unrealistic) initial condition of the population to be $\delta(0)$ distributed at the beginning of its spread, one must look for the movement equation of the process that will generate the spatiotemporal distribution of the population for any initial distribution. Since both temporal and spatial dimensions are involved, the movement equation must be a partial differential equation in this case, with a left-hand side expressing the rate of change for $\hat{p}(x, t)$ at any spatiotemporal point (x, t). This rate is the first-order time derivative of $\hat{p}(x, t)$ as given by (2.15), that is,

$$\frac{\partial \hat{p}(x, t)}{\partial t} = \frac{(x^2 - 2Dt) \cdot \exp\left(-\dfrac{x^2}{4Dt}\right)}{8Dt^2 \sqrt{Dt\pi}}. \tag{2.19}$$

It is reasonable to expect the local rate of density change at a point x to be determined by the inhomogeneity of the spatial distribution of population density around that point. If the spatial distribution is homogeneous, all of its derivatives with respect to x are of a constant zero value everywhere. On the other hand, spatial inhomogeneity can be characterized by non-zero spatial derivatives of order one or greater. Comparing the space derivatives to (2.19), the time derivative of the spatiotemporal distribution, it is the second-order derivative of $\hat{p}(x, t)$ with respect to x that turns out to be closest to (2.19), being

$$\frac{\partial^2 \hat{p}(x, t)}{\partial x^2} = \frac{(x^2 - 2Dt) \cdot \exp\left(-\dfrac{x^2}{4Dt}\right)}{8D^2 t^2 \sqrt{Dt\pi}}. \tag{2.20}$$

On dividing (2.19) by (2.20) and rearranging, we end up with the simple partial differential equation

$$\frac{\partial \hat{p}(x, t)}{\partial t} = D \cdot \frac{\partial^2 \hat{p}(x, t)}{\partial x^2}. \tag{2.21}$$

Equation (2.21) is called the diffusion equation, of which (2.15) is a particular solution. The diffusion equation was first formulated more than a century ago by a physiologist, A. Fick, who studied the diffusion of colloidal particles in a viscous medium. In fact it was his work that initiated theoretical research on the diffusion processes.

On multiplying (2.21) by N, the total number of individuals of the population, we get

$$\frac{\partial p(x, t)}{\partial t} = D \cdot \frac{\partial^2 p(x, t)}{\partial x^2} \quad \text{with} \quad p(x, t) = \hat{p}(x, t) \cdot N, \tag{2.22}$$

a diffusion equation that regards directly $p(x, t)$, the density of the population (that is, the **number** of individuals per unit of space), instead of $\hat{p}(x, t)$,

24   Reaction–diffusion models

the probability density (or the **fraction** of population per unit of space). Hence (2.22) relates the rate of population density change at a given point to the steepness of the population density gradient at that point. Note that $p(x, t)$ is not a PDF any more, since its integral over the whole infinite $(-\infty, \infty)$ range of x equals N, not one. It is just this property that makes (2.22) more convenient than (2.21) for modelling population diffusion processes. The constraint of total probability conservation would be violated anyway with the demography of the population incorporated and thus a source term appearing in diffusion models (Sections 2.5 and 2.6) or else the information on total population abundance would be lost. Therefore it is better to give up the rigorous probabilistic interpretation related to (2.21) now, and refer to (2.22) by the name 'diffusion equation' henceforth.

2.3.1   The flow balance approximation to diffusion

The diffusion equation (2.22) can also be deduced directly relying upon general population density flow balance considerations, without obtaining any particular solution first (Okubo, 1980). Let us start once more from the model of random walk in an infinite one-dimensional habitat. We postulate that the individuals move according to the same rules as before, except that now we allow for anisotropic movement as well. In each time unit of duration τ, all individuals move once to a distance of λ, either to the right or to the left, but the probability of the two directions may be different; say it is α to the right, and $\beta = 1 - \alpha$ to the left. We would like to gain a general formula that gives the population density at the spatiotemporal point (x, t) for any value of x and t. With these conditions given, the simple relation

$$p(x, t) = \alpha \cdot p(x - \lambda, t - \tau) + \beta \cdot p(x + \lambda, t - \tau) \qquad (2.23)$$

holds, stating that the population density in the spatial interval of length dx centred at point x in time t is composed of the density influx from the intervals (of length dx) at $(x - \lambda)$ and $(x + \lambda)$, according to the above specified rules of anisotropic random walk. Since each individual moves with certainty in each time unit, no individuals that were at x in time $t - \tau$ remain there by time t, therefore the right-hand side of (2.23) does not contain terms of $p(x, t - \tau)$, and $\alpha + \beta = 1$ holds. We denote the difference of the step probabilities in the two directions by $\varepsilon \equiv \alpha - \beta$, so that $\varepsilon = 0$ represents isotropic movement with both directions equally probable. Now assume that $p(x, t)$ is a sufficiently smooth function in both time and space, that is, the population density within a small interval of length λ of the habitat is large enough to exclude random fluctuations in density both between neighbouring intervals at a given point of time and between adjacent time units at a given site. If this requirement is met, then the function p can be replaced by its Taylor expansion (for a short introduction to the Taylor expansion of functions and the actual procedure for $p(x, t)$, see

Appendix A) around (x, t) for the points $p(x - \lambda, t - \tau)$ and $p(x + \lambda, t - \tau)$. This reads

$$p(x-\lambda, t-\tau) = p(x, t) - \tau \frac{\partial p(x, t)}{\partial t} - \lambda \frac{\partial p(x, t)}{\partial x} + \frac{\tau^2}{2} \cdot \frac{\partial^2 p(x, t)}{\partial t^2}$$
$$+ \tau\lambda \cdot \frac{\partial^2 p(x, t)}{\partial t \partial x} + \frac{\lambda^2}{2} \cdot \frac{\partial^2 p(x, t)}{\partial x^2} + \ldots$$
$$p(x+\lambda, t-\tau) = p(x, t) - \tau \frac{\partial p(x, t)}{\partial t} + \lambda \frac{\partial p(x, t)}{\partial x} + \frac{\tau^2}{2} \cdot \frac{\partial^2 p(x, t)}{\partial t^2}$$
$$- \tau\lambda \cdot \frac{\partial^2 p(x, t)}{\partial t \partial x} + \frac{\lambda^2}{2} \cdot \frac{\partial^2 p(x, t)}{\partial x^2} + \ldots$$
(2.24)

Substituting (2.24) into (2.23), applying $\alpha + \beta = 1$ and $\alpha - \beta = \varepsilon$ and rearranging terms yields

$$\frac{\partial p(x, t)}{\partial t} = -\frac{\lambda\varepsilon}{\tau} \cdot \frac{\partial p(x, t)}{\partial x} + \frac{\lambda^2}{2\tau} \cdot \frac{\partial^2 p(x, t)}{\partial x^2} + \lambda\varepsilon \cdot \frac{\partial^2 p(x, t)}{\partial t \partial x}$$
$$+ \frac{\tau}{2} \cdot \frac{\partial^2 p(x, t)}{\partial t^2} + \ldots$$
(2.25)

Since we wish to get a partial differential equation for $p(x, t)$ with the same constraints on the parameters λ, τ and ε as those set for (2.22), we let λ and τ decrease through all limits but require λ^2/τ to approach $2D$ at the same time, and assume isotropic movement with $\varepsilon = 0$ (we return to the case with $\varepsilon \neq 0$ later, cf. Section 2.4.1). Then all terms except for the second one vanish on the right-hand side of (2.25), so that

$$\frac{\partial p(x, t)}{\partial t} = D \cdot \frac{\partial^2 p(x, t)}{\partial x^2} \quad \text{with} \quad D = \frac{\lambda^2}{2\tau}, \tag{2.26}$$

which is (2.22), the model of Fickian diffusion.

Note that we did not state anything in this section about the initial distribution of the population, neither did we make use of the properties of any specific PDF to arrive at (2.26); therefore it is valid for any reasonable (that is, relatively smooth) initial distributions. Equation (2.15) is a single particular solution of (2.26), for the cases of $\delta(0)$ or Gaussian initial distributions. It can be shown, however, that the conclusion of the linear time dependence of variance at the rate $2D$ holds for all solutions of the diffusion equation, regardless of the initial distribution, provided the spread of the population is spatially unlimited.

2.3.2 Density flow

From the assumptions defining the isotropic random walk process, one can directly obtain the flow of population density through a given point x of the habitat, applying a third type of approximation to the diffusion process. Consider two adjacent intervals of length λ, with their boundary at x. We suppose the density within the intervals to be concentrated in the centres of the intervals, that is, at $x - \lambda/2$ and $x + \lambda/2$. This assumption is unrealistic as it stands, but it will cause no problems later, since we will let $\lambda \to 0$. Each of the $\lambda \{p(x - \lambda/2, t)\}$ individuals within the left interval moves to one of the neighbouring intervals with a probability dt/τ in a short time $dt < \tau$, that is, they leave it in any one of the two possible directions with a probability $dt/2\tau$ each (because of the isotropy of the walk). The same applies to the $\lambda\{p(x + \lambda/2, t)\}$ individuals in the right interval. Therefore the net flow of population density during dt through the boundary at x in the direction left to right, expressed as the difference between the flow in the positive and the negative direction, respectively, is

$$J(x, t) \cdot dt = \frac{\lambda}{2\tau} \{p(x - \lambda/2, t) - p(x + \lambda/2, t)\} dt, \tag{2.27}$$

where $J(x, t)$ is the **flux density**, that is, the number of individuals per unit time passing through point x in time t (Figure 2.6). We use the Taylor expansion in x of the right-hand side of (2.27) around (x, t) the same way as we did for (2.23), and again let $\lambda \to 0$ and $\tau \to 0$ so that $\lambda^2/2\tau \to D$, which results in

$$J(x, t) = -D \frac{\partial p(x, t)}{\partial x}. \tag{2.28}$$

Equation (2.28) says that the flux density through the point x at time t is proportional to the density gradient $\partial p/\partial x$ at x, and the constant diffusivity D. The negative sign indicates that the net flux is positive through x (a net flow of density occurs from left to right) if the density gradient is negative around x (the density decreases left to right). By comparing (2.22) and (2.28),

$$\frac{\partial p(x, t)}{\partial t} = -\frac{\partial J(x, t)}{\partial x} \tag{2.29}$$

follows, which can be shown to be of general validity, regardless of the mechanism of particle movement defined. That (2.29) is a general form can be easily seen intuitively, if one considers that the only thing it states is the simple fact that the density change at point x within a small interval dx after

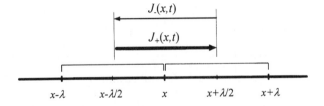

$$J(x,t) = J_+(x,t) - J_-(x,t)$$

Figure 2.6 Population density flow through the spatial point x at time t. Net population density flux through x is $J(x, t)$.

a short time dt is the **net efflux** of density from the interval dx (the difference of the flux densities through the right and left borders of the interval dx), **whatever microscopic mechanism** we assume to determine the actual density flux $J(x, t)$. For isotropic random walk, that is, in Fickian diffusion, $J(x, t)$ turns out to be (2.28), but for other mechanisms it may take completely different forms – and (2.29) remains valid in any case. We shall rely heavily on this fact later.

2.3.3 Diffusion in 2D and 3D spaces

Based on both the random walk model and the flow balance model, the diffusion equation can be extended to account for population dispersion in more than one spatial dimensions. This is also an essential requirement if one intends to make the diffusion approach more realistic, since most of the habitats of actual populations are far from being one dimensional. There are important exceptions, however, like long, narrow rivers, lakes or canyons, in which cases an appropriate one-dimensional model might be a sufficiently accurate approximation to describe population dispersion. Apart from these, most actual terrestrial and aquatic habitats are two or three dimensional (fractal dimensions are not mentioned here; we shall return to this topic later; cf. Section 5.3.6).

The simplest possible assumption to start with is again that the movement of individuals is isotropic, meaning that the speed and the mechanism of dispersion is independent of spatial direction. First we see how this can be realized in the random walk model and the resulting diffusion equation in two dimensions. Imagine an area divided by a rectangular grid with a cell side length of λ, and an individual released in the focal cell centred at ($x_0 = 0$, $y_0 = 0$). Let the individual move randomly within the grid so that in a time unit of duration τ it steps into one of the four **diagonally** neighbouring cells (that is, exactly one step is taken in each of the spatial dimensions within a

28   Reaction–diffusion models

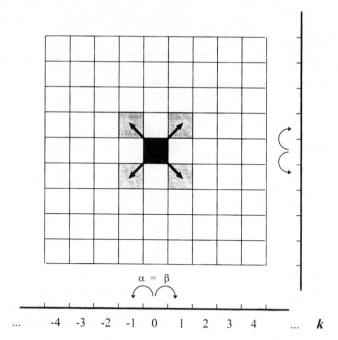

Figure 2.7 Isotropic random walk in two discrete spatial dimensions, as composed of two independent one-dimensional walks.

time unit). The four diagonal directions are equally likely to be taken, with a probability of 0.25, which implies that the projections of the movement onto the two dimensions are independent of each other (Figure 2.7). Moreover, with the current assumptions, each of the projected movements is identical to the one-dimensional random walk described in detail earlier, regarding both its mechanism and its parameters, λ and τ. This allows us to make use of the formalism applied to the one-dimensional case, with only a few notational modifications.

Starting directly with the continuous case, we look for the (bivariate) PDF of the movement in two dimensions, from which it is possible to determine the probability of the presence of the individual within any section of the (x, y) plane. Relying on the assumption that the movements in the two dimensions are independent, we can compose the joint PDF from its one-dimensional projections by simple multiplication as

$$\hat{p}(x, y, t) = \hat{p}(x, ., t) \cdot \hat{p}(., y, t), \tag{2.30}$$

where $\hat{p}(x, ., t)$ and $\hat{p}(., y, t)$ are the PDFs for the projected component movements in the x and y dimensions, respectively, and $\hat{p}(x, y, t)$ is the two dimensional PDF. With

$$\hat{p}(x, ., t) = \frac{1}{\sqrt{4\pi Dt}} \cdot \exp\left(-\frac{x^2}{4Dt}\right)$$

and (2.31)

$$\hat{p}(., y, t) = \frac{1}{\sqrt{4\pi Dt}} \cdot \exp\left(-\frac{y^2}{4Dt}\right)$$

we have

$$\hat{p}(x, y, t) = \frac{1}{4\pi Dt} \cdot \exp\left(-\frac{x^2 + y^2}{4Dt}\right). \quad (2.32)$$

Equation (2.32) is the PDF of the bivariate Gaussian distribution with an expected value of $\mathbf{m} = \mathbf{0}$ and $\sigma^2 = 2Dt$ variance (Figure 2.8). It is a particular solution of a two-dimensional partial differential equation, which may be extracted from it by direct differentiation much like in the one-dimensional case (cf. (2.19)–(2.21)), except that now we have to produce second-order derivatives with respect to both x and y of the right-hand side to get

$$\frac{\partial \hat{p}(x, y, t)}{\partial t} = D \cdot \left(\frac{\partial^2 \hat{p}(x, y, t)}{\partial x^2} + \frac{\partial^2 \hat{p}(x, y, t)}{\partial y^2} \right) = D \cdot \nabla^2 \hat{p}(x, y, t). \quad (2.33)$$

∇^2 denotes the scalar operation of adding second derivatives in x and y. Multiplying (2.33) by N, the total number of individuals, the resulting equation specifies $p(x, y, t)$, the population density at the spatiotemporal point (x, y, t):

$$\frac{\partial p(x, y, t)}{\partial t} = D \cdot \nabla^2 p(x, y, t), \quad \text{with} \quad p(x, y, t) = N \cdot \hat{p}(x, y, t). \quad (2.34)$$

The extension of the approach to three spatial dimensions can be carried out along a similar line of reasoning with the same kind of formal manipulations, and it leads to a form perfectly analogous with (2.34).

2.3.4 Methodological overview: initial and boundary conditions, solutions, stability analyses and numerical approximations of PDE models

Before going on to set up and to analyse actual reaction–diffusion models, a brief summary of the general principles and methods of diffusion modelling might be useful. This section stresses a few important methodological points, some of which will be used or referred to in concrete models later in this chapter.

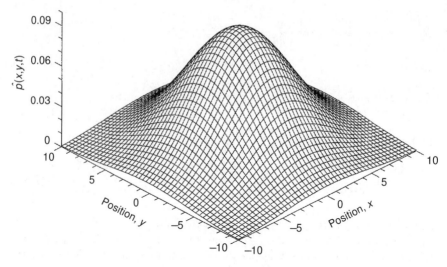

Figure 2.8 A time slice of the spatiotemporal PDF of an isotropic, two-dimensional, continuous random walk process.

For a diffusion problem to be completely specified, and thus for the related model to be a hopeful candidate for being investigated by analytical means, it is necessary to specify the partial differential equation governing the dynamics, but this is not in itself sufficient. One has to postulate on at least two other important conditions for that.

1. Is the habitat of the population bounded? If so, what happens to the individuals who reach the boundaries?
2. What was the initial distribution $p(x, 0)$ of the population at time $t = 0$ within the boundaries of the habitat?

Answering these questions in a proper form means setting up the **boundary conditions** and the **initial condition** of the model, respectively.

Boundary conditions can be of several types, depending on the model situation to be explored. Ricciardi (1985) reports a formal classification of boundaries originally given by Feller (1952, 1954), but for our present purposes it will be sufficient to mention just three characteristic types on a more practical basis. A boundary is defined as a set of points delimiting a finite interval of space (which may be one, two or three dimensional). We denote the set of boundary points by B, and apply the notational convention of vector calculus, that is, boldface letters denote vector variables and functions. Thus, **x** may be a point in a one- to three-dimensional space, $p(\mathbf{x}, t)$ is the population density at that point in time t, and $\mathbf{J}(\mathbf{x}, t)$ is the flux density vector at point **x** in time t. We shall use the following types of boundary conditions.

1. Suppose we confine our interest to a finite section of a practically infinite topographical region, but in fact nothing special happens to those individuals who reach the boundaries of the area of interest (Figure 2.9a). Then we have an open system with a **passive** boundary delimiting a 'window' through which a part of the system is observed. The choice of a passive boundary may be justified for example in the case of a study regarding a regional subpopulation not naturally isolated from the remaining part of the population. In fact the points of a passive boundary do not represent singular places regarding the dispersion mechanism, therefore no formal conditions different from those specified by the model equation are necessary. However, the mathematically convenient but otherwise harmless assumption that at points infinitely far from the origin both the density p and the flux density $\mathbf{J}$ of the population vanish is usually made.

2. If the trajectory of the movement of individuals ends at a finite distance from the origin so that they disappear from the habitat immediately after having encountered the edge, the boundary is **absorbing**. It is specified by the boundary condition

$$p(\mathbf{x}, t) = 0, \quad \text{for} \quad \mathbf{x} \in B, \qquad (2.35)$$

stating that the population density is always zero at the boundary (Figure 2.9b). Clearly an absorbing boundary is not a realistic assumption for most actual animal population dynamical problems, since it implies the immediate death or removal of all individuals arriving at B. Therefore, if actively moving organisms are considered, who are able to detect the dangerous edge before being killed or caught, it should not be used in spite of the fact that it may simplify the formalism a great deal (Nisbet and Gurney, 1982). However, such a boundary condition may be fair enough for phytoplanktonic organisms or seed-dispersed plants in isolated habitats with penetrable (but deadly) edges, supposing that individuals or seeds crossing the boundary are lost forever.

3. If the boundary of an isolated habitat is impenetrable for the members of a population so that they have to turn back soon after they reach it, then the boundary is a **reflecting** one (Figure 2.9c). The appropriate formal condition to be incorporated into the models of such situations should express the fact that at each point of B the component of the flux density $\mathbf{J}$ parallel to the norm of B at that point is 0. A formal condition conforming to this and only to this requirement would be rather complicated to formulate and use, but there is another, which includes this condition as a special case, yet it is convenient mathematically. We assume that all components of $\mathbf{J}$ are 0 at B, that is,

$$\mathbf{J}(\mathbf{x}, t) = 0 \quad \text{for} \quad \mathbf{x} \in B. \qquad (2.36)$$

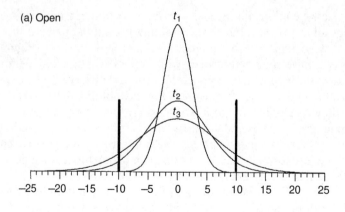
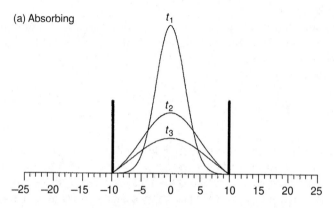
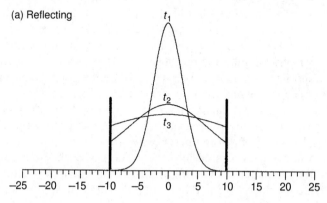

Figure 2.9 Solutions of the diffusion equation (2.21) with different boundary conditions: (a) no active boundaries; (b) absorbing boundaries; and (c) reflecting boundaries.

The diffusion equation 33

By adopting (2.36) we do not lose too much generality, because the only kind of solution it excludes is the quite implausible outcome of a vortical flow of individuals along the whole length of the boundary.

Islands, habitat islands, fenced experimental plots, lakes and vessels of bacterial cultures are only some of the innumerable examples of closed 'habitats' which may justify the use of the reflecting boundary condition. It is therefore not surprising that most case studies based on diffusion models apply (2.36) as the boundary condition.

Absorbing and reflecting boundaries are **active boundaries** in the sense that they alter the behaviour of the individuals when encountered. On the boundaries the validity of the model equation is suspended, and local rules are defined. The interval of space within which the model equation governs the process is termed the **diffusion range**.

The initial distribution $p(\mathbf{x}, 0)$ can be set up arbitrarily within the diffusion range, but it should not be forgotten that different initial population density patterns may occasionally lead to qualitatively different results. The initial condition of the spatial distribution of the population at $t = 0$ as a multiple of the Dirac-δ or of a normal PDF is often a prerequisite for certain kinds of diffusion models to be tractable analytically (Ricciardi, 1985). This may be a serious limitation, if one tries to use actual field distributions as initial conditions of a diffusion model.

Once we have a partial differential equation governing a certain diffusion process in hand, with the boundary condition and the initial distribution specified, we wish to use it for prediction, that is, we try to forecast how the population will be distributed in space for any point of time t, or at least to say what the limiting distribution is like as $t \to \infty$. To explore, in all possible details, the dynamics of a certain model, one has to solve the differential equation, that is, the explication of the function $p(\mathbf{x}, t)$, specifying population density for any spatiotemporal coordinates. This involves a series of formal manipulations on the original PDE, including integration over both time and space, which is not necessarily possible for any diffusion model; in fact, models with known solutions are exceptions rather than the rule. If a general solution exists, it is still quite a troublesome task to find it; it can require a great deal of mathematical invention and much patience, even if the model equation itself seems very simple and innocent.

An alternative way of extracting information on the dynamics of a diffusion model is to find its equilibrium distribution (or distributions) $p^*(\mathbf{x})$, if there is such, and then ask about the stability properties of $p^*(\mathbf{x})$. All methods of stability analysis are partial as compared to explicit solution, since they yield no information about the trajectories ending at the equilibrium distribution(s). In equilibrium distribution, the rate of population density change $dp^*(\mathbf{x})/dt$ is 0 at all spatial points, under the given boundary conditions. If the equilibrium distribution is not trivial (that is, $p^*(\mathbf{x}) \neq 0$ for at least some $\mathbf{x}$), we can ask about its stability, that is, what happens if

34   Reaction–diffusion models

$p^*(\mathbf{x}, t)$ is perturbed, for example by adding or removing individuals, so that the density change at time t_0 at $\mathbf{x}$ is of a value $\zeta(\mathbf{x}, t_0)$. The problem is to determine the direction of change for the perturbation $\zeta(\mathbf{x}, t)$, as $t \to \infty$. If it vanishes everywhere, then the density distribution returns to its shape before perturbation, and it is said to be asymptotically stable; if $\zeta(\mathbf{x}, t)$ does not change with time, the distribution is neutrally stable; if $\zeta(\mathbf{x}, t)$ increases, the distribution is unstable. Which of these possibilities occurs can be determined by different methods, the most frequently applied ones being local linearization and the Liapunov test. Local linearization is a relatively simple matter, but the Liapunov test is usually very complicated, if it is possible at all. The stability analysis of a stationary distribution might involve even more demanding mathematical tools, such as Fourier transforms and convolutions. Apart from local linearization, we shall not go into the details of these procedures.

If all else fails, there are still a number of numerical methods that give useful information on how the system in question behaves. Numerical methods are completely computer-oriented; the principle is to discretize the model equations in both time and space, applying one of the many algorithms available, and then use the resulting difference equations to generate an approximate graph of the density distribution for any time step. A numerical solution is always particular in at least two respects. First, it can encompass only a finite – often very small – slice of the spatiotemporal continuum. Second, it gives an outcome for a specific set of model parameters; other sets of parameter values produce different results. Therefore, if one needs to have a relatively general view on what dynamics the model follows within a certain interval of its parameter space, one has to run the program many times with many different parameter sets within that interval. How accurate a view we can get by these methods therefore depends largely on the capacity of the computer and software used.

2.3.5   Solutions of the diffusion equation

Now let us return to the specific model of Fickian diffusion. When starting from the isotropic random walk approximation which led to (2.15) – a particular solution of (2.21) – without actually having to solve it, we have at the outset automatically specified both the initial and the boundary conditions, by postulating that

1. the habitat is of infinite topographical extension, that is, no active boundaries are imposed on the system, and
2. the population was released at a single point, $x_0 = 0$, that is, the initial distribution was the δ function centred at $x = 0$.

For the diffusion equation (2.22) and its two- and three-dimensional extensions, the analytical solution is known for any initial distribution. It is composed as a weighted superposition of an infinite number of fundamental ((2.15) type) solutions; the weighting factor is the actual initial density $p(x, 0)$ in each spatial point x. For the one-dimensional version, the general solution looks like

$$p(x, t) = \frac{1}{\sqrt{4\pi t}} \cdot \int_{-\infty}^{\infty} p(\xi, 0) \cdot \exp\left(-\frac{(x-\xi^2)}{4Dt}\right) d\xi. \tag{2.37}$$

The following important dynamical properties of (2.37) can be shown (see Levin (1976, 1985b) for details):

1. the total number of individuals, N does not change with time, since, as it can be shown by integration,

$$\int_{-\infty}^{\infty} p(x, t) dx = \int_{-\infty}^{\infty} p(x, 0) dx = N; \tag{2.38}$$

2. $p(x, t)$ approaches 0 as $t \to \infty$, which is obvious by inspecting (2.37): the first term containing t vanishes, whereas the exponential term approaches one as t passes to the limit;
3. the abundance centre of the population, defined as the expected value M of the distribution,

$$M = \frac{\int_{-\infty}^{\infty} x \cdot p(x, t) dx}{N}, \tag{2.39}$$

   does not move with t;
4. the variance $V(t)$ of x,

$$V(t) = \frac{\int_{-\infty}^{\infty} (x-M)^2 \cdot p(x, t) dx}{N}, \tag{2.40}$$

   is a linearly increasing function in t: $V(t) = V(0) + 2Dt$.

All these properties are straightforward intuitively from the postulates of the random walk approximation that led to the diffusion equation (2.22) (in fact, point 1 is one of the explicit basic assumptions in the random walk model). The two- and three-dimensional extensions of this model yield the same conclusions, which can be interpreted as follows.

The simplest possible diffusion model for population dispersion predicts that a large population of identical non-multiplying individuals who are moving at random, and do not react to any possible inhomogeneities or heterogeneities in the milieu by changing their simple pattern of movement, will disperse in a habitat of infinite extension according to the model of Fickian diffusion. If no active boundaries of the habitat exist, then the variance of their spatial distribution is expected to increase with time in a linear fashion, whereas the density centre of the population remains static.

What if an active boundary is imposed on this system? If the boundary is absorbing, then we expect the whole population to flow out of the diffusion region through the boundaries and thus go extinct ultimately as $t \to \infty$. With reflecting boundaries assumed, the population tends to be evenly distributed within the diffusion range, regardless of what the initial distribution was. It is easy to see this for the one-dimensional system within the reflecting boundaries $r_1 < r_2$. The equilibrium density distribution can be found from

$$\frac{dp^*(x)}{dt} = 0 = D \cdot \frac{d^2 p^*(x)}{dx^2} \quad \text{and} \quad J^*(x) = -D\frac{dp^*(x)}{dx} = 0 \qquad (2.41)$$

with $\dfrac{\partial p(r_i, t)}{\partial x} = 0, \quad i = 1, 2.$

It is obvious that the equilibrium flux density $J^*(x)$ can be zero everywhere only if the population is evenly distributed, that is, if $p^*(x) = \text{constant}$, supposing $D \neq 0$. Hence, given the boundary conditions and the initial distribution $p(x, 0)$,

$$p^*(x) = \frac{\int_{r_1}^{r_2} p(x, 0) dx}{r_2 - r_1} = \frac{N}{r_2 - r_1}. \qquad (2.42)$$

As a consequence, the abundance centre M^* of the population approaches the mass centre of the diffusion range, and the variance V^* of x tends towards a finite value,

$$M^* = (r_1 + r_2)/2 \quad \text{and} \quad V^* = \int_{r_1}^{r_2} (x - M^*)^2 \cdot \frac{1}{r_2 - r_1} dx = \frac{(r_2 - r_1)^2}{12}. \qquad (2.43)$$

Note that changing the boundary conditions altered the dynamical behaviour qualitatively, as compared to that without an active boundary: for absorbing boundaries it is only conclusion 2 above that still holds for any initial condition, while for reflecting boundaries, the only conclusion remaining valid is 1.

2.3.6 Density-dependent diffusion with biased random walk

Moving in a habitat like a Brownian particle in a fluid medium does not require much more ability from an individual than what a relatively isometric molecule possesses. To improve our assumption of this somewhat dull behaviour, we may suppose that the individuals are able to detect the density of the population in their immediate neighbourhood, and adjust the frequency or the length of their steps accordingly. If crowding is regarded as disadvantageous by the individual, it can try to escape by simply stepping more or longer than before. The resulting movement is an isotropic, **biased** random walk.

The adjustment of step frequency or step length requires some kind of perception of local density from individuals, and ability to act accordingly, properties we usually attribute to animals, but not to plants. Even terrestrial plants can sometimes meet this requirement, however, if they are able to control their frequency of offspring production, or the position of allocated offspring, according to local density. This is quite often the case with vegetatively propagating plants (Harper, 1977), but it might be true for seed-dispersed species as well, with seed production and the morphology of 'dispersal facilities' like hairs on the seeds adjusted according to local density stress. In such cases, density-dependent diffusion modelling may be applicable to plant populations as well.

The bias of random walk can be incorporated into the diffusion equation by assuming that the diffusivity D is not a constant, but a function of local density $p(x, t)$, that is, $D[p(x, t)]$. The simplest possible form of this kind defines a linear dependence of D on p, so that

$$D[p(x, t)] = D_0 + D_1 \cdot p(x, t). \tag{2.44}$$

However, the constant D cannot be simply replaced by $D[p(x, t)]$ in the diffusion equation, since the diffusion equation (2.22) is a specific form of the general equation (2.29), with $J(x, t)$ specified for the unbiased random walk approximation. We have to go back to (2.29), and find the specific form of $J(x, t)$ for the biased case. This can be done, for example, by reconsidering (2.27), and assuming that the frequency or the length of steps increases with local density. We shall derive the model for the case of increased step frequency below. The simplest convenient assumption is that the probability of stepping within a short time unit dt increases linearly with local population density $p(x, t)$, that is

$$1/\tau = a + b \cdot p(x, t). \tag{2.45}$$

Then (2.27) is replaced by

$$J(x, t)\mathrm{d}t = \frac{\lambda}{2}\{a + b \cdot p(x, t)\}\{p(x - \lambda/2, t) - p(x + \lambda/2, t)\}\mathrm{d}t \tag{2.46}$$

After Taylor expansion, and taking the limits as $\lambda \to 0$, $\tau \to 0$, $1/a \to 0$ and $1/b \to 0$ so that $\lambda^2 a/2 \to D_0$ and $\lambda^2 b/2 \to D_1$, we end up with

$$J(x, t) = -\left(D_0 + D_1 \cdot p(x, t)\right) \cdot \frac{\partial p(x, t)}{\partial x}. \tag{2.47}$$

It is (2.47) that we substitute into (2.29), to obtain

$$\begin{aligned}\frac{\partial p(x, t)}{\partial t} &= -\frac{\partial}{\partial x}\left(-\left(D_0 + D_1 p(x, t)\right)\frac{\partial p(x, t)}{\partial x}\right) \\ &= D_0 \frac{\partial^2 p(x, t)}{\partial x^2} + D_1\left(\left(\frac{\partial p(x, t)}{\partial x}\right)^2 + p(x, t)\frac{\partial^2 p(x, t)}{\partial x^2}\right)\end{aligned} \tag{2.48}$$

Equation (2.48) consists of a density-independent and a density-dependent component of diffusion. $D_0 + D_1 \cdot p(x, t)$ is called the **virtual diffusivity** of the population. For certain applications, such as modelling the movement of animals seeking territory in a favourable habitat, it may be straightforward to require D_0 to be 0, which results from setting $a = 0$ in (2.45) (no diffusive movement in the absence of conspecific competitors). Nisbet and Gurney (1982) call the corresponding movement **directed**, albeit this name is somewhat misleading: it is not the direction, but the intensity of the movement that is affected by density in this case.

If we suppose that in addition to density-dependent diffusion, that there is some inhomogeneity in the external milieu affecting diffusibility, then the constants D_0 and D_1 of the virtual diffusivity are not constants any more, but functions of spatial position. In this case, (2.48) is modified to

$$\frac{\partial p(x, t)}{\partial t} = \frac{\partial^2}{\partial x^2}\left\{\left[D_0(x) + D_1(x) \cdot p(x, t)\right] \cdot p(x, t)\right\}. \tag{2.49}$$

The right-hand side of (2.49) could be further explicated according to the rules of differentiation, but the result would be very long, and even less instructive than the simple density-dependent system (2.48). Shigesada and Teramoto (1978) analyse this model in detail.

2.4 ADVECTION

The assumption that the individuals of a population move in a perfectly isotropic fashion is another unrealistic idealization inherent in (2.22), which is temporarily justifiable by mathematical convenience, but it has to be relaxed soon for diffusion models to be of more theoretical, let alone practical value in population dynamics. Anisotropic movement can be defined in terms of the underlying random walk process by stating that either (a) step probability depends on direction or (b) step length depends on direction (Figure 2.10). From an interpretative point of view, the two cases may be the results of different causes. Supposing that it is the individual who decides which direction to take, preference towards an attractive centrum, or aversion from a repelling locality may be expressed by altering the probability of stepping in the corresponding direction. The same may apply to adjusting step lengths according to a preferred direction, that is, stepping longer towards the attractive object, or away from the repelling one. Such motivating objects and the 'intention' of the individual need not play a role for differential step length adjustment, however, since the same result comes from supposing a steady flow by speed u of the medium, if the individual walks in an isotropic fashion relative to the medium. The speed of its movement relative to the fixed coordinate system is then increased in the 'downstream' direction and decreased 'upstream' by the speed u.

2.4.1 Constant rate advection with step probability adjustment

For the anisotropic step probability case, consider (2.23), and its Taylor-expanded form (2.25) with $\alpha \neq \beta$, that is, $\varepsilon \neq 0$. The only terms in (2.25) not vanishing in this case as τ and λ approach zero are the first two on the right-hand side. However, if ε is held constant, the first term grows infinitely, since the speed of walk $\lambda/\tau \to \infty$ in the limit, provided $\lambda^2/\tau \to 2D$. Therefore we let $\varepsilon \to 0$ as $\lambda \to 0$ and $\tau \to 0$ so that $\lambda\varepsilon/\tau \to u$ at the same time, and get

$$\frac{\partial p(x,t)}{\partial t} = -u\frac{\partial p(x,t)}{\partial x} + D\frac{\partial^2 p(x,t)}{\partial x^2}. \tag{2.50}$$

The only acceptable reason for taking these limits is again a mathematical one, fitting to the constraint of the speed of walk increasing to infinity in the limit. Equation (2.50) differs from the diffusion equation (2.22) in the first term of the right-hand side, which is called the **advection term**, and accounts for a directed flow of population density with a constant advection velocity u along the x dimension. That this is a correct interpretation of the advection term can be easily seen if one considers the flux density $J(x,t)$ as composed of a directional (advective) flow component and a diffusive component. The directional component should be such that in a short time interval dt all individuals within the habitat interval $(x, x - u\cdot dt)$ should flow through the point x. Therefore, supposing the interval $u\cdot dt$ is short enough for $p(x,t)$ to be considered as constant there, $J(x,t)dt = u\cdot p(x,t)dt$ without diffusion. If both advective and diffusive flows are considered, the two should be superimposed, so that

40 Reaction–diffusion models

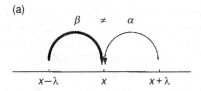

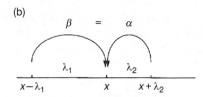

Figure 2.10 Two realizations of anisotropic random walk: (a) anisotropic step probabilities; and (b) anisotropic step lengths.

$$J(x, t) = up(x, t) - D\frac{\partial p(x, t)}{\partial x}. \tag{2.51}$$

Now recall the general form (2.29),

$$\frac{\partial p(x, t)}{\partial t} = -\frac{\partial J(x, t)}{\partial x} = -u\frac{\partial p(x, t)}{\partial x} + D\frac{\partial^2 p(x, t)}{\partial x^2}, \tag{2.52}$$

which is (2.50).

2.4.2 Constant rate advection with step length adjustment

Now let us turn to the case with step lengths depending on direction (Figure 2.10b). Assume that the preferred direction is the positive one (left to right), so that $\lambda_1/\lambda_2 = 1 + \mu$, and the probability of stepping to any direction is 0.5, that is, $\varepsilon = 0$. Then (2.23) modifies to

$$p(x, t) = \frac{1}{2}p(x - \lambda_1, t - \tau) + \frac{1}{2}p(x + \lambda_2, t - \tau). \tag{2.53}$$

Taylor expansion and rearrangement in the same way as for (2.23) leads to

$$\frac{\partial p(x, t)}{\partial t} = \frac{\lambda_2 - \lambda_1}{2\tau}\frac{\partial p(x, t)}{\partial x} + \frac{\lambda_1^2 + \lambda_2^2}{4\tau}\frac{\partial^2 p(x, t)}{\partial x^2} + \ldots \tag{2.54}$$

Now let $\lambda_2 \equiv \lambda$ that is, $\lambda_1 = \lambda \cdot (1 + \mu)$, and substitute these into (2.54):

$$\frac{\partial p(x,t)}{\partial t} = -\frac{\lambda\mu}{2\tau}\frac{\partial p(x,t)}{\partial x} + \left(\frac{2\lambda^2}{4\tau} + \frac{2\lambda^2\mu}{4\tau} + \frac{\lambda^2\mu^2}{4\tau}\right)\frac{\partial^2 p(x,t)}{\partial x^2} + \ldots \quad (2.55)$$

Taking the limits $\lambda, \tau, \mu \to 0$ so that $\lambda^2/\tau \to 2D$ and $\lambda\mu/\tau \to 2u$, we arrive at (2.50) again. The conclusion is that with the assumption of a constant anisotropy either in step probability or in step length, we end up with the same diffusion–advection model (2.50) with constant coefficients u and D.

The solution of (2.50) is known for the general $p(x, 0)$ initial distribution, without active boundaries:

$$p(x,t) = \frac{1}{\sqrt{4\pi t}} \cdot \int_{-\infty}^{\infty} p(\xi, 0) \exp\left(-\frac{(x-\xi-ut)^2}{4Dt}\right) d\xi. \quad (2.56)$$

Equation (2.56) is the same distribution as (2.37), the one for simple diffusion, except that its mean shifts at a constant speed u away from the origin. The variance of the distribution is the same as for the isotropic case (2.40); it increases with time linearly. Assuming the origin to be a reflecting boundary and the initial distribution to be the delta function $N \cdot \delta(x_0)$ ($x_0 > 0$), the solution is also known, but it is too long and complicated to be reproduced here (for the precise form, see Ricciardi, 1985). As one may expect, if there is only one reflecting boundary at the origin, and u is positive (that is, if the mean of the density distribution shifts away from the origin), the effect of reflection diminishes with time, and the abundance distribution flattens out through all limits, like it does without active boundaries. If, on the other hand, u is negative, the system admits a stable density distribution which is sloping down from the origin, because advection causes density to accumulate around the reflecting boundary. How quickly the distribution converges to equilibrium, depends on D and u.

Helland et al. (1989) apply a diffusion–advection model to explain the data gathered in a release–recapture experiment (Helland et al., 1984) with bark beetles. Actual recapture, as measured by the number of beetles caught in a pheromone trap, fitted very well to the predicted values in all the cases reported. Note that advection was induced by an attractive centre (the source of sex pheromone), and not a flow of the medium in this experiment. This is one of the most successful applications of diffusion modelling in explaining a specific field situation.

2.4.3 Advection induced by a milieu gradient

Advection velocity is of course not necessarily constant. It can, for example, depend on spatial position, if the anisotropy of step probability (ε) or that of step length (μ) depends on x somehow. An example of such a spatial dependence often encountered in the literature is an advective density flow along the gradient $\partial\psi/\partial x$ of a certain abiotic milieu factor, e.g., nutrient (upward advection, towards higher concentrations) or toxic (downward advection towards lower concentrations) substances. In such a case, advection velocity u can be defined as

$$u(x) = u_0 \frac{\partial \psi(x)}{\partial x} \qquad (2.57)$$

if the milieu factor has a steady spatial distribution; otherwise advection velocity becomes also time-dependent:

$$u(x, t) = u_0 \frac{\partial \psi(x, t)}{\partial x}. \qquad (2.58)$$

Negative u_0 means downward advection (the milieu factor is repelling, $u(x)$ is positive if the gradient is negative), whereas positive u_0 values represent upward advection (the milieu factor is attractive, $u(x)$ is negative for negative gradient).

With (2.58) in effect, (2.50) is modified to

$$\begin{aligned}\frac{\partial p(x, t)}{\partial t} &= -\frac{\partial J(x, t)}{\partial x} = -\frac{\partial}{\partial x}\left(u_0 \frac{\partial \psi(x, t)}{\partial x} p(x, t) - D \frac{\partial p(x, t)}{\partial x}\right) \\ &= -u_0 \left(\frac{\partial^2 \psi(x, t)}{\partial x^2} p(x, t) + \frac{\partial \psi(x, t)}{\partial x} \frac{\partial p(x, t)}{\partial x}\right) + D \frac{\partial^2 p(x, t)}{\partial x^2},\end{aligned} \qquad (2.59)$$

which describes linear diffusion and advection towards (or against) the concentration (intensity) gradient of a milieu factor. An application of milieu-dependent advective terms will be considered in Section 2.6.3, model (2.84).

2.5 REACTION I: POPULATION GROWTH IN DIFFUSIVE SYSTEMS

In principle at least, almost the complete subject of classical population dynamics can be incorporated into diffusion models under the title 'reaction', with a spatiotemporal reinterpretation of well-known concepts such as population growth, density dependence, competition, parasitism, predation, etc. The term 'reaction' covers all these, and many more population dynamical processes – those related to the temporal change in abundance. The word itself is obviously inherited from chemistry, where it means processes altering the concentrations of compounds through chemical interactions that create and destroy molecules.

Without reaction terms introduced, diffusion models are of limited interest from a population dynamical point of view, because neither births, nor deaths, nor population interactions (that is, reactions in the strict sense of the word) occur in a purely dispersing population. What diffusion and advection in themselves can produce is simply the redistribution of the population within the habitat, while its abundance remains static globally. Thinking in terms of biological populations, the assumption that no demographic events occur on the time scale justifying a diffusion approximation is itself contradictory in most actual field cases. There are exceptions to this, exemplified by some field experiments with insect populations whose spatial dispersion was followed on a time scale excluding reproduction (e.g., Skellam, 1951, 1973; Helland et al., 1989).

Once we have the (2.22) diffusion model for the dispersion of a population formulated, it is straightforward to extend it by including demographic terms. These should be defined in a spatiotemporal setting, that is, as functions of both time and space. As before, we shall follow the bottom-up strategy of introducing more complex dynamical relations into the diffusion–advection model, starting with the simplest possible case of exponential population growth for a single population, then incorporating density dependence and interaction with other populations. This section is devoted to models of within-population reactions, whereas Section 2.6 deals with species interactions in diffusive systems.

2.5.1 Constant rate growth and dispersion

To derive the model of exponential population growth combined with anisotropic dispersal, we start from the model of anisotropic random walk (2.23). Assume that each individual dies with a probability v within a time unit of length τ. This means that deaths reduce local density p by a factor $(1-v)$ during a time unit. If an individual survives, it produces an average of η offspring during τ, which means that local population density increases by a factor $(1+\eta)$. Thus the proportion of population density (relative to the density at the beginning of the time unit) that will be dispersed into the neighbourhood by the end of the time unit is $(1+\eta)(1-v) \equiv (1+\gamma)$. The probabilities of stepping right and left are α and β, respectively, so that $(\alpha + \beta) = 1$, and $(\alpha - \beta) \equiv \varepsilon$, as before. With these conditions, (2.23) is modified to

$$p(x, t) = \alpha(1+\gamma)p(x-\lambda, t-\tau) + \beta(1+\gamma)p(x+\lambda, t-\tau). \tag{2.60}$$

Taylor expansion and rearrangement yields

$$\frac{\partial p(x, t)}{\partial t} = \frac{\gamma}{(1+\gamma)\tau} \cdot p(x, t) - \frac{\lambda\varepsilon}{\tau} \cdot \frac{\partial p(x, t)}{\partial x} + \frac{\lambda^2}{2\tau} \cdot \frac{\partial^2 p(x, t)}{\partial x^2} + \ldots \tag{2.61}$$

Now let $\lambda \to 0$, $\tau \to 0$, $\varepsilon \to 0$, $\gamma \to 0$ so that $\lambda^2/\tau \to 2D$, $\lambda\varepsilon/\tau \to u$, and $\gamma/\tau \to g$. Then all but the first three terms on the right-hand side of (2.61) vanish, and what remains can be written as

$$\frac{\partial p(x,\,t)}{\partial t} = gp(x,\,t) - u\frac{\partial p(x,\,t)}{\partial x} + D\frac{\partial^2 p(x,\,t)}{\partial x^2}. \tag{2.62}$$

Equation (2.62) is the simplest possible diffusion–advection–reaction system, with $g \cdot p(x, t)$, the **reaction** or **source term** of the model being linear in $p(x, t)$; g is the spatiotemporal equivalent of the intrinsic rate of increase in the Malthusian model of exponential population growth. Recalling the chemical analogy: if g, u and D are positive, (2.62) describes the kinetics of an autocatalytic reaction in a laminar flow reactor. In terms of population dynamics, it is a model for an exponentially growing and diffusively dispersing population in a water or air current.

The right-hand side of (2.62) is a linear combination of the derivatives of $p(x, t)$ of order zero to two with respect to x, and the coefficients are the rate constants g, $-u$ and D. If the population does not disperse, that is, if the advection and the diffusion terms are of zero value, (2.62) degenerates to the Malthusian model of exponential population growth. This may happen either if both coefficients u and D are zero, or if the density gradient $\partial p(x, t)/\partial x$ is zero everywhere, which means overall spatial homogeneity. Then the abundance dynamics of the population can be represented by the solution of the simple Malthusian ordinary differential equation $dN(t)/dt = gN(t)$, which is

$$N(t) = N(0) \cdot \exp(g \cdot t). \tag{2.63}$$

If this is the model for the demography of a homogeneously distributed population, what would it be for one which is inhomogeneously distributed in space? The simple answer is: just the same, provided the boundary of the habitat is not absorbing. If the rate of local density change g is not dependent on x, either directly or indirectly through some other parameters or variables of the model – themselves functions of x – population growth and dispersion happen to be independent of each other. The solution of (2.62) consists of two independent parts, one for population growth, the other for dispersion. With the general initial distribution $p(x, 0)$ and no active boundaries, the solution is

$$p(x,\,t) = e^{gt} \cdot \frac{1}{4\pi t} \int_{-\infty}^{\infty} p(\xi,\,0) \exp\left(-\frac{(x - \zeta - ut)^2}{4Dt}\right) d\xi, \tag{2.64}$$

where e^{gt} is the growth factor, and the remaining part of the right-hand side is the same as (2.56), that is, the distribution resulting from diffusion and advection. The same applies to any diffusion model with a constant growth rate, irrespective of the dispersion mechanism, if the boundaries are not absorbing. Of course if $g > 0$, the population grows, and with $g < 0$ it shrinks to extinction exponentially. Hill (1990) and Davis et al. (1991) use model (2.62) with negative growth rates in mortality studies of drifting and dispersing larval populations of lobsters and tuna fish. It is essentially the same model, but without advection, that Liddle et al. (1987) use for estimating the loss rate and the diffusion coefficient of a seed population, making use of the fact that growth and dispersion are disconnected when the growth term is exponential. Gerrodette (1981) estimated also the growth rate g and the diffusivity D of a solitary coral population with a model of the same kind.

It is obvious that $p(x, t)$ is always positive in (2.64) for all finite values of x, if t is positive and no active boundaries are assumed, even if the initial distribution was $p(x, 0) = \delta(0)$. This is an artefactual characteristic of the basic model (2.22) of diffusion, arising from the unrealistic assumption of an infinite speed of random walk in the limit (see Section 2.2.2). This nonsense feature, which is inherited by each model with a constant-rate diffusion component, can be improved by supposing a threshold density below which the population is not detectable. If the initial distribution was $\delta(0)$, the wave front of detectable density advances with a constant speed depending on r, D, t, the initial and the threshold density. Andow et al. (1990) use this modified model in an attempt to explain the observed speed of dispersion for some animal species' populations (muskrat in Europe, cereal leaf beetle and small cabbage butterfly in North America). They found quite a good accordance of theoretical predictions to observed data.

2.5.2 The critical habitat size problem

With an absorbing boundary condition defined, even the simple exponential reaction–diffusion model

$$\frac{\partial p(x, t)}{\partial t} = g \cdot p(x, t) + D \cdot \frac{\partial^2 p(x, t)}{\partial x^2} \qquad (2.65)$$

becomes very interesting from the point of view of the persistence of a dispersing population. Suppose that the absorbing boundaries are at a distance L from each other. If L is small and D, the diffusivity, is relatively large, we expect an intensive population density flow out of the diffusion range. In such a case the population can persist or grow within the boundaries only if g, the growth rate is sufficiently large to compensate for the diffusive losses. With L relatively long compared to diffusivity, a small positive growth rate might be enough to maintain a viable population (a positive effective growth rate) within the boundaries. As first demonstrated by Kierstead and Slobodkin (1953) in an attempt to explain the characteristic patch size of oceanic phytoplankton blooms, (2.65) admits a positive net growth only if a simple condition relating the model constants g, D and the length of the diffusion range, L holds, namely if

$$L > L_c = \pi\sqrt{D/g}. \tag{2.66}$$

L_c is the critical habitat length (for a one-dimensional habitat), below which the population goes extinct even if g, the instantaneous local growth rate is itself positive. A similar relation can be obtained for two- and three-dimensional isodiametric patches, with only the coefficient changing slightly, within the same order of magnitude. In two dimensions, for example, π should be replaced by the value 4.81, but (2.66) remains valid otherwise.

2.5.3 Density-dependent growth and dispersion

As a necessary next step towards the biological relevance of diffusion modelling, one has to consider rules for population growth more realistic than the Malthusian one, since exponential population increase is a temporary phenomenon in nature, bound to a massive excess of depletable resources as compared to demands. When the quantity of any of these resources decreases to a level where it becomes limiting, competition among the individuals of the population begins for that resource, slowing down or even stopping further increase in abundance. It is straightforward to assume that local competitive pressure increases with local density, which in turn reduces the growth rate of the population at the given locality. To take this effect into account, any one of the classical forms for density-dependent population growth can be applied with a local interpretation. Following Nisbet and Gurney (1982), we consider two of these in some detail: the Gompertz form and linear density dependence (the latter resulting in a spatial version of the well-known logistic model).

The Gompertz form for the local density dependence of the growth rate is

$$g(p(x, t)) = -c \cdot \ln\frac{p(x, t)}{K}, \tag{2.67}$$

where c and K are positive constants. It is clear by inspecting (2.67) that the constant K (called the local carrying capacity) is the population density at which no population growth occurs: $g(K) = 0$. The spatiotemporal points (x, t) where the local density is below K may be regarded as **sources** with a positive net population growth, and points where the density exceeds K are **sinks** with negative growth rates (Pulliam, 1988). It is also seen from (2.67) that the growth rate approaches infinity as local density decreases, which is not a realistic assumption, thinking in terms of the inevitable physiological constraints on the maximum rate of offspring production. This problem can be overcome by shifting the function left with a constant value a so that

$$g(p(x, t)) = -c \cdot \ln\frac{a + p(x, t)}{K}, \tag{2.68}$$

in which case the maximum growth rate at $p(x, t) = 0$ is finite: $g(0) = -c\ln(a/K)$, and the carrying capacity is $K - a$. Nisbet and Gurney (1982) apply this modified form as the growth term in a reaction–diffusion model of a density-limited population within absorbing boundaries. The model equation is

$$\frac{\partial p(x, t)}{\partial t} = -c\ln\frac{a + p(x, t)}{K}\cdot p(x, t) + D\frac{\partial^2 p(x, t)}{\partial x^2}, \qquad (2.69)$$

which, being non-linear in the growth term, cannot be solved, only its stability properties can be analysed. If the habitat size L (that is, the length of the diffusion interval, or the distance of the boundaries) exceeds the critical value $L_c = \pi(D/l)^{\frac{1}{2}}$, the population would grow exponentially, if no density dependence were included. One would expect that the critical habitat size L_c' should be larger for a self-limiting population than for one which can grow exponentially regardless of its density, but eventually this turns out not to be the case. The existence of a non-trivial stable equilibrium density distribution $p^*(x)$ for (2.69) can be proved applying the Liapunov method, if the relation $L > L_c$ holds, which is the same criterion as in the exponential, density-independent model. The heuristic explanation of this is straightforward: if the habitat length exceeds L_c by only a very small value, the stationary distribution of the exponential model is very flat; the local densities are close to zero everywhere. If local densities are small, introducing density dependence plays no significant role in the dynamics: $p^*(x)$ gets sufficiently close to zero, if L gets close enough to L_c. Thus, the critical viable habitat size is not affected by the density dependence introduced, but it stops the growth of population density ultimately, however large the value of L. The steady-state distribution is, of course, non-uniform, since $p^*(x) = 0$ on the boundaries, and $p^*(x) > 0$ within the diffusion interval. These results are also valid for the two-dimensional extension of (2.69).

If the growth rate is linear in $p(x, t)$ with a negative slope, the resulting quadratic growth term of the reaction–diffusion system (2.70) is a spatiotemporal equivalent of the classical logistic model:

$$\frac{\partial p(x, t)}{\partial t} = r_0\left(1 - \frac{p(x, t)}{K}\right)\cdot p(x, t) + D\frac{\partial^2 p(x, t)}{\partial x^2}, \qquad (2.70)$$

where r_0 is the growth rate at very low densities and K is the carrying capacity; both are positive constants. This model has been intensively studied since Fisher proposed it more than half a century ago (Fisher, 1937) as a possible model for the dispersion of advantageous alleles in a population. Closed form solutions are not known to this model either, for the same reason as for (2.69), but some of its interesting properties have been investigated under different boundary conditions.

If there is no active boundary on the system, it can be shown that a $\delta(0)$

distributed initial population, governed by (2.70), will grow and disperse in such a way that after a sufficiently long time has elapsed, a wavefront of population density propagates at a constant speed c in each direction, with its shape unchanged, as shown on Figure 2.11. Based on the ultimate stationarity of the shape of the advancing wavefront, that is, on the relation $p(x, t_0 + t) = p(x - ct, t_0)$, it can be shown that

$$c = 2\sqrt{r_0 D}, \qquad (2.71)$$

that is, the speed of the wavefront increases with both the growth rate and the diffusivity of the population. For the mathematical details of the analysis, the reader is referred to Skellam (1973).

Clearly the critical patch size problem is also pertinent for the logistic reaction–diffusion model (2.70) with absorbing boundaries. Relying on his

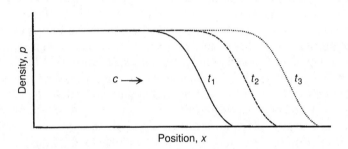

Figure 2.11 Advancing wavefronts: travelling wave solutions of (2.70) (density-dependent growth and passive diffusion). c is the ultimate speed of the wavefront as $t \to \infty$.

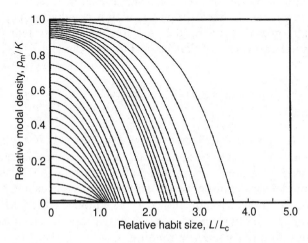

Figure 2.12 Shapes of the spatial density distributions for different relative habitat sizes (L/Lc), in absorbing boundary solutions of (2.70) (after Okubo (1980)).

proof of the existence of a stable unimodal steady-state distribution, Skellam (1951) found a relation between the habitat length normalized by the minimum viable habitat size for the exponential model, L/L_c, and the modal density of the steady-state distribution relative to the carrying capacity, p_m/K. The relation can be calculated in analytical form, as a complicated equation (it can be found, e.g., in Skellam (1951) and Okubo (1980)), but it is more instructive to observe the graph of the relation instead of the algebraic form (Figure 2.12). The graph shows that – if the habitat size L exceeds L_c, the minimum viable habitat length of the exponential model – p_m is always positive, and it approaches K in the limit as L tends to infinity. The minimum length of habitat necessary for a self-limiting and dispersing population to survive is the same as if it were growing exponentially. This result accords to that of Nisbet and Gurney (1982) obtained for the model (2.69) with the Gompertz form of density dependence; the heuristic explanation is also the same.

2.6 REACTION II: SPECIES INTERACTIONS IN DIFFUSIVE SYSTEMS

A completely homogeneous environment is quite an exceptional situation in which hardly any natural population finds itself for a long time. Even if the relevant abiotic factors of the milieu can be safely assumed to be uniformly distributed in space, which is in itself quite improbable, it is almost certain that the biotic milieu component, consisting of effects due to the presence of one or more potentially interacting species within the habitat, will show spatiotemporal fluctuations. The environmental inhomogeneities thus generated follow the abundance patterns of the interacting populations, so that they become mutually dependent on each other's spatial pattern.

In fact this is the point of the complexity of reaction–diffusion models from which the outcome of a specific model cannot usually be foreseen by pure intuition. In many cases, multispecies diffusion models produce surprising, even counter-intuitive results; therefore the formal treatment of such problems is indispensable if one wishes to get a qualitative view, however crude, on the dynamical behaviour of the system in question. Unfortunately, however, this is also the point at which analytical treatment becomes very difficult for most models; numerical methods become of prevalent importance here, either as means of generating hypotheses which can be then rigorously proven or rejected by analytical techniques, or as the only practicable method of extracting information from a complicated system.

2.6.1 The Lotka–Volterra diffusion model

Interactions can be logically built into diffusion models by extending the density dependence concept to interspecific relations, so that the growth rate of each population become a function of the density of all others. The

50   Reaction–diffusion models

resulting model embodies an equation for each species (and sometimes additional equations for abiotic factors depending on population densities), thus forming a system of coupled partial differential equations. The biotic interactions can, of course, be any of the familiar types: predation, competition, mutualism, etc. The form in which these are incorporated into the reaction terms of the model equations is usually a close relative of a non-spatial model, e.g., of the Lotka–Volterra type.

In a non-spatial model, the only possible interaction among the species is demographic: they can mutually affect the numbers of births and deaths in each other's populations. In a reaction–diffusion model, there is a second type of interaction to be specified, affecting directly the mechanism of spatial dispersion. It is a plausible assumption that the presence of different species may influence the movement pattern of the individuals differentially, depending, for example, on the type of the demographic interaction. Predators chase their prey (prey attracts predators), and prey try to escape from the predator (predator repels prey), to take just a trivial example. Such differential dispersal interactions can be modelled by the inclusion of so-called **cross-diffusion** (Kerner, 1959; Jorné, 1977) terms in reaction–diffusion models. Cross-diffusion means diffusive movement determined by the density gradient of another species' population. To construct the simplest (linear) cross-diffusion term specifying the effect of species j on the dispersion of species i, it is most convenient to start from the density-flow argumentation applied in Section 2.3.2 in the derivation of (2.28), the linear formula of autodiffusion. We assume that the jth component of the flux density of species i through the point x is determined by the density gradient of species j at x:

$$J_{ij}(x, t) = -D_{ij}\frac{\partial p_j(x, t)}{\partial x}. \tag{2.72}$$

Assuming that the flux density components are independent of each other, the total flux density of species i can be calculated by simple addition as

$$J_i(x, t) = -\sum_{j=1}^{s} D_{ij}\frac{\partial p_j(x, t)}{\partial x}. \tag{2.73}$$

Mutatis mutandis applying (2.29), the general formula of density flow, we get

$$\frac{\partial p_i(x, t)}{\partial t} = -\frac{\partial J_i(x, t)}{\partial x} = \sum_{j=1}^{s} D_{ij}\frac{\partial^2 p_j(x, t)}{\partial x^2}, \tag{2.74}$$

which is a generalization of the original diffusion equation (2.22) in the sense that (2.22) can be regained from (2.74) by simply letting the cross-diffusivities $D_{ij} = 0$ for all $j \neq i$. Unlike the autodiffusion coefficients D_{ii},

Population growth in diffusive systems 51

which are always non-negative, cross-diffusivities can be positive, negative or zero, depending on the dispersal interaction of the corresponding pair of species. A negative cross-diffusivity means the movement of i individuals towards higher densities of species j, that is, chasing of species j by species i. The generalized diffusitivities D_{ij} can be arranged in the **diffusivity matrix**, which is usually asymmetric, and contains autodiffusivities in its diagonal and cross-diffusivities elsewhere.

A simple Lotka–Volterra diffusion model in one-dimensional space, for s interacting and dispersing populations, with constant advection velocities and constant cross-diffusivities is

$$\frac{\partial p_i}{\partial t} = r_i \left(1 + \frac{\sum_{j=1}^{n} a_{ij} p_j}{K_i(x, t)}\right) p_i - u_i \frac{\partial p_i}{\partial x} + \sum_{j=1}^{s} D_{ij} \frac{\partial^2 p_i}{\partial x^2}, \quad i = 1, \ldots, s. \quad (2.75)$$

The reaction term of (2.75) is a Lotka–Volterra-type growth function, containing $r_i > 0$, the growth rate of species i without intra- and interspecific interaction (corresponding to the physiological maximum rate of growth), $K_i(x, t)$, the carrying capacity of the environment for species i at the spatiotemporal point (x, t), and a_{ij}, the interaction coefficients measuring the effect of species j on species i. Positive a_{ij} values mean beneficial interaction (e.g., the effect of the presence of prey on predator, or mutualistic benefit), whereas negative a_{ij} values represent adverse effects (e.g., predator effect on prey, competition). Of course, $a_{ij} = 0$ represents neutrality, that is, no interaction. u_i is the advection velocity of species i, and D_{ij} is the cross-diffusivity of species i with respect to species j, the only constraint being $D_{ii} \geq 0$ for all i.

Many of the diffusion models with population interactions studied so far are special cases of (2.75). Steele (1974), Murray (1975), Rosen (1975), Mimura and Nishida (1978), Dunbar (1983) and many others analysed, for example, the simplest possible predator–prey diffusion model, formulated as the pair of partial differential equations

$$\frac{\partial H}{\partial t} = aH - bHP + D_H \frac{\partial^2 H}{\partial x^2}$$

$$\frac{\partial P}{\partial t} = -cP + dHP + D_P \frac{\partial^2 P}{\partial x^2}, \quad (2.76)$$

where $H(x, t)$ stands for the density of the prey (herbivore) and $P(x, t)$ for the density of the predator population; the coefficients are positive constants. The reaction terms are the spatial analogues of those in the classical Lotka–Volterra model of predation, which has a non-trivial, neutrally stable

52   Reaction–diffusion models

equilibrium point at $P^* = a/b$, $H^* = c/d$. Thus, if the system is open or the boundaries are reflecting, and the initial distributions of the populations are both uniform, the diffusion terms vanish, and (2.76) uniformly oscillates around the (H^*, P^*) equilibrium state everywhere. If the equilibrium is uniformly perturbed, that is, if H^* is increased by a constant value ζ_H and P^* by ζ_P everywhere (infinite wavelength perturbation), then diffusion will play no role in determining the behaviour of the system, since the density gradients remain zero everywhere after perturbation. This means that the densities of both populations will oscillate infinitely about H^* and P^* in a spatially uniform manner, provided the boundary is not absorbing. The model with a spatially uniform initial distribution reproduces the behaviour of the non-spatial Lotka–Volterra system of predation.

Equation (2.76) behaves very different with different initial and boundary conditions, however. Non-uniform (that is, finite wavelength) perturbations damp and spread out to infinity in open systems, if $D_P/D_H = 1$. Without active boundaries imposed, and with equal diffusivities for prey and predator, the system (2.76) is asymptotically stable against local perturbations. If $D_P/D_H > 1$, the perturbation ultimately spreads with a constant speed as a travelling wavefront, followed by a train of waves of decreasing amplitude (Chow and Tam, 1976), and it may last for a long time to damp the perturbation even near the site it appeared. With absorbing boundaries, local perturbations dissipate from the diffusion range through the 'leaking' margins of the habitat. Reflecting boundaries, on the other hand, retain local perturbations within the diffusion range, but they tend to 'smooth out' by time, which eventually results in a uniform oscillation of population density about the uniform equilibrium with a period of $2\pi(ac)^{-\frac{1}{2}}$. In conclusion, neither open nor bounded systems governed by (2.76) can produce stable non-uniform spatial patterns in the limit as $t \to \infty$.

2.6.2 Diffusive instability in models of interacting species

Considering the possible role of diffusive-reactive morphogenes in the control of embryonic tissue differentiation, Turing (1952) proved that if two reactive substances have different diffusivities, stable non-uniform concentration patterns of standing waves may emerge from a uniform initial distribution, as a result of local perturbations (see also Meinhardt (1982) and Murray (1989)). The ecological reinterpretation of the phenomenon called **diffusive instability** or **Turing effect** is due to Segel and Jackson (1972), Levin (1974, 1977), Levin and Segel (1976), Okubo (1978), McMurtrie (1978), Kishimoto (1982) and others. It was Segel and Jackson (1972) who gave a simple necessary and sufficient condition for the Turing effect to appear in the two-species interacting diffusion system (2.77), based on the stability analysis by local linearization of a general model with linear autodiffusion. We consider this model in some detail, since it can be applied

Species interactions in diffusive systems

for many specific two-species cases, and the mathematics required is rather simple.

The model equations are

$$\frac{\partial S_1}{\partial t} = F_1(S_1, S_2) + D_1 \frac{\partial^2 S_1}{\partial x^2}$$
$$\frac{\partial S_2}{\partial t} = F_2(S_1, S_2) + D_2 \frac{\partial^2 S_2}{\partial x^2},$$
(2.77)

where the reaction terms F_i are functions of the densities S_i of the interacting populations such that they allow for the existence of a spatially uniform equilibrium (S_1^*, S_2^*), which is stable against uniform perturbations. If we set

$$S_i(x, t) = S_i^* + \zeta_i(x, t),$$
(2.78)

(the actual abundance at spatiotemporal coordinates (x, t) are expressed as the sum of the local equilibrium value and the perturbation term ζ_i) and substitute it into (2.77), we get

$$\frac{\partial \left(S_i^* + \zeta_i(x, t) \right)}{\partial t} = F_i\left(\left[S_i^* + \zeta_i(x, t) \right], \left[S_j^* + \zeta_j(x, t) \right] \right) + D_i \frac{\partial^2 \left(S_i^* + \zeta_i(x, t) \right)}{\partial x^2}$$
(2.79)

where $i, j \in \{1, 2\}, i \neq j$.

Taylor expansion of F_i about (S_1^*, S_2^*) and rearrangement of the other terms yields

$$\frac{\partial S_i^*}{\partial t} + \frac{\partial \zeta_i(x, t)}{\partial t} = F_i\left(S_1^*, S_2^* \right) + \frac{\partial F_i\left(S_1^*, S_2^* \right)}{\partial S_i} \zeta_i(x, t)$$
$$+ \frac{\partial F_i\left(S_1^*, S_2^* \right)}{\partial S_j} \zeta_j(x, t) + \ldots + D_i \frac{\partial^2 S_i^*}{\partial x^2}$$
$$+ D_i \frac{\partial^2 \zeta_i(x, t)}{\partial x^2},$$
(2.80)

where $i, j \in \{1, 2\}, i \neq j$.

Being close to the equilibrium as the perturbation $\zeta(x, t)$ is small, we retain only the linear terms of the Taylor series. Recall that (S_1^*, S_2^*) is an equilibrium, hence $F_i(S_1^*, S_2^*) = 0$, $\partial S_i^*/\partial t = 0$, and that this equilibrium is spatially uniform with $\partial S_i^*/\partial x = 0$. Then we are left with

$$\frac{\partial \zeta_1(x, t)}{\partial t} = a_{11} \cdot \zeta_1(x, t) + a_{12} \cdot \zeta_2(x, t) + D_1 \frac{\partial^2 \zeta_1(x, t)}{\partial x^2}$$
$$\frac{\partial \zeta_2(x, t)}{\partial t} = a_{21} \cdot \zeta_1(x, t) + a_{22} \cdot \zeta_2(x, t) + D_2 \frac{\partial^2 \zeta_2(x, t)}{\partial x^2}. \quad (2.81)$$

where $a_{ij} = \dfrac{\partial F_i\left(S_1^*, S_2^*\right)}{\partial S_j}$, $i, j \in \{1, 2\}$

Equation (2.81) is a linear diffusion equation in the perturbations $\zeta_i(x, t)$. We enquire as to the asymptotic behaviour of its solutions as $t \to \infty$. If $\zeta_i(x, t)$ approaches zero everywhere for both species, then the uniform equilibrium distribution is stable against small perturbations, that is, model (2.77) with the actual reaction terms permits no Turing effect.

Segel and Jackson (1972) found the necessary and sufficient condition for (2.77) to admit diffusive instability to be

$$a_{11}D_2 + a_{22}D_1 > 2\sqrt{D_1 D_2 \cdot (a_{11}a_{22} - a_{12}a_{21})} > 0. \quad (2.82)$$

By simply evaluating $\partial F_i/\partial S_j$ ($i, j \in \{1, 2\}$), the Jacobian matrix of F (called the 'community matrix' in ecological context (Levins, 1968)) at the equilibrium point, we can decide about the existence of diffusive instability for any specific form of (2.77). Equation (2.82) implies a number of simple necessary conditions, which may serve as a 'first filter' in determining for a specific model whether it admits the Turing effect or not (Okubo, 1980). Such relations are

[a] $a_{11}a_{22} < 0$ and $a_{12}a_{21} < 0$
[b] if $a_{ii} < 0$ then $D_i > D_j$ $\quad (2.83)$

From (2.83a) it is obvious that both populations must be dependent on their own densities, one of them positively (self-accelerator or destabilizer), the other negatively (self-regulator or stabilizer). For example, this requirement excludes (2.76), the basic type of Lotka–Volterra predatory interactions with linear autodiffusion discussed above from among the models with a possible Turing effect, since for the reaction term of (2.76), $a_{11} = a_{22} = 0$. It is also seen by inspecting (2.83a) that mutual interactions like competition (mutual disadvantages) or symbiosis (mutual benefits) can not produce

Species interactions in diffusive systems

diffusive instability in two-species systems, if the species disperse by linear diffusion (Evans, 1980). Equation (2.83b) says that the stabilizer population must disperse better than the destabilizer. Equation (2.82) applies to two-species systems only. Mimura (1984) shows that competitive systems with the number of species above two may exhibit diffusive instability. Kishimoto (1982) demonstrated the same for competitive and predatory systems with three species. Takeuchi (1991) examined a three-species competition model and showed that diffusion can induce regional stability in cyclic competitive interaction systems which are unstable without diffusion.

Okubo (1980) reviews a number of specific models of the type (2.77), in the search for criteria for the occurrence of diffusive instability. A brief synopsis of their biologically most relevant conclusions is as follows. The Lotka–Volterra system for a self-accelerating phytoplanktonic and a self-regulating zooplanktonic grazer population admits diffusive instability with the diffusivities properly chosen (Levin and Segel, 1976). So too does a similar model with the functional response and the self-regulation of the predator incorporated (Levin, 1977). A system consisting of a limiting nutrient with a constant supply rate, a phytoplanktonic population that depends on that resource and a phytophagous animal grazing on it at a constant rate also shows diffusive instability within a small section of its parameter space (Okubo, 1978). A generalized Lotka–Volterra model for predation with density dependence for both populations, and Allee effect for the prey, may produce the Turing effect, if the equilibrium density of the prey is under the Allee threshold (Mimura and Murray, 1978). Jorné (1974, 1975, 1977), Rosen (1977), Mimura and Kawasaki (1980) and Namba (1989) demonstrated that the inclusion of appropriate terms for advection and/or cross-diffusion tends to promote diffusive instability in closed systems (those with reflecting boundaries), if they act so as to keep the interacting populations apart in space (see Holmes *et al.* (1994) for a review). Murray (1989) provides a detailed study of diffusion-driven spatial instability in relation to single- and multispecies ecological interactions; the interested reader is referred to this book for a more detailed mathematical treatment of the subject.

2.6.3 Competitive coexistence through habitat-partitioning

As mentioned earlier, the violation of the criteria in (2.81) exclude the possibility of pattern development by diffusive instability for competitive systems of the type (2.77). Shigesada *et al.* (1979) shows, however, that if the linear mechanism of dispersion is replaced with both density- and milieu-dependent diffusion and milieu-dependent advection, then competing populations can coexist by partitioning the habitat even if the underlying non-spatial model predicts the competitive exclusion of one of the competitors. Diffusive instability can be a result of non-linear advection and diffusion terms. Their two-species example was based on the numerical analysis of the pair of partial differential equations

56   Reaction–diffusion models

$$\frac{\partial S_i(x,\,t)}{\partial t} = \frac{\partial^2}{\partial x^2}\left[D_i(x,\,S_1,\,S_2)\cdot S_i(x,\,t)\right] + \gamma_i\frac{\partial}{\partial x}\left[\frac{\partial\phi(x)}{\partial x}S_i(x,\,t)\right]$$
$$+ R_i(S_1,\,S_2)$$

for $(i,\,j \in \{1,\,2\},\ i \neq j)$,

where $R_i(S_1,\,S_2) = (r_i - a_{ii}S_i(x,\,t) - a_{ij}S_j(x,\,t))\cdot S_i(x,\,t)$
$D_i(x,\,S_1,\,S_2) = \{\alpha_i(x) + \beta_{ii}(x)S_i(x,\,t) + \beta_{ij}S_j(x,\,t)\}$

(2.84)

The first term on the right-hand side is the diffusion term, with $D_i(x, S_1, S_2)$, the virtual diffusivity of species i, being a two-species extension of that included in (2.49). The second term represents milieu-dependent advection up the intensity gradient $\partial\phi(x)/\partial x$ of an attractive environmental factor (that is, advection velocity is the same as (2.57), with $u_i(x) = \gamma_i \cdot \partial\phi(x)/\partial x$). $R_i(S_1, S_2)$ is the reaction term, which prescribes local population growth according to the Lotka–Volterra model for two-species competition. The reaction term alone is known to define a system admitting either the coexistence of both competitors, or the exclusion of one, depending on simple criteria regarding the values of the constants r_i and a_{ij}. Exclusion may be unconditional (which of the two competitors survives is independent of their initial abundances)

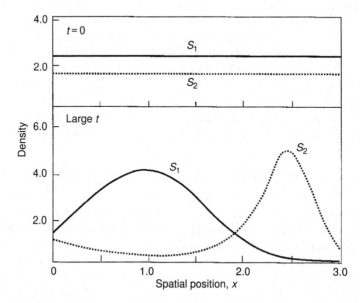

Figure 2.13 The coexistence of competitors through habitat partitioning, induced by diffusive instability in the diffusion–reaction–advection model (2.84) (after Shigesada *et al.* (1979)).

or conditional (the outcome depends on initial abundances). Since analytical methods are impracticable on (2.84), Shigesada *et al.* (1979) integrated it numerically. They found that with the parameters of the model properly chosen, the system admits coexistence even if the reaction term alone would allow only for conditional exclusion. Diffusion and advection acted together to induce regional coexistence. The constants γ_1 and γ_2 in the advection term were chosen to be close, that is, the affinities of the two species towards the attractive environmental factor were similar. The spatial distribution of the competitors after a long elapsed time is shown in Figure 2.13, from which we see that the coexistence of the competitors occurs through the spatial division of the habitat, in the form of spatially separate standing waves, due to the specific mechanism of non-linear diffusion and advection.

2.7 SUMMARY

When searching for methods to treat spatial aspects of population interactions in a theoretical framework, one reasonable option is to extend the analogy (or, in fact, the homology) between the models of chemical kinetics and population dynamics, by including diffusion mechanisms into population models. The multitude of reaction–diffusion systems proposed in population dynamics, mainly in the past four decades, have come a long way from being simple analogues of the physicochemical models of reaction kinetics: there are quite general concepts of the reaction–diffusion approach that originated from population biological models (e.g., cross-diffusion, density-dependent diffusion, etc.).

The main assumptions behind the partial differential equation formulation of reaction–diffusion models are different from those of classical, non-spatial population dynamics in two important respect. First, all relevant population interactions are local, that is, the classical assumption of overall environmental homogeneity and perfect spatial mixing can be relaxed. The vital attributes of the populations and the environmental variables may be explicitly dependent on spatial position either directly or indirectly, via the local abundances. Second, besides the usual types of demographic interactions in non-spatial models, diffusion models can account for dispersal interaction, that is, for density- and milieu-dependent effects modifying the dispersion movement of the individuals.

Questions pertaining to persistence and coexistence problems are inseparable from those concerning the spatial patterns generated in reaction–diffusion systems. Even relatively simple non-linear reaction–diffusion–advection systems are capable of diffusive instability, producing static or dynamic mesoscopic patterns from a spatially homogeneous initial state. Static patterns are standing waves, whereas dynamic patterns can be advancing or even rotating waves of population abundance. Diffusive instability might induce regional coexistence, even if the corresponding non-spatial population model predicts exclusion – the immediate reason of coexistence is always the emergence of a mesoscale pattern and the local separation of the interacting populations thereby.

By the inclusion of spatial dimensions, the dynamics of interacting populations can be studied on many different scales. As a result, concepts like persistence, coexistence or dispersal become scale-dependent, just as they are in any actual field situation. The increased biological realism of the reaction–diffusion approach, as compared to that of non-spatial modelling, is reflected in the scattered published examples of field-tested reaction–diffusion systems.

3 Population dynamics in patchy environments

3.1 INTRODUCTION

The spatial range of even a very common species is rarely if ever continuous on all spatial scales; it is usually a patchwork of more or less isolated habitats within a matrix of hostile environmental conditions, resulting either from natural heterogeneities of the environment, or from landscape-transforming human activity. The patchy distribution of species abundance seems to be the rule rather than the exception – and it has important consequences regarding the dynamics of interacting populations. For this reason, **patchy environment** problems are of much practical and theoretical interest, which motivated the vast literature of patch modelling in population and community dynamics.

'Patchy environment' situations are found on very different spatio-temporal scales, the most obvious one being 'species on oceanic islands' – but the concept and the related formalism apply very well to a wide class of ecologically very different field and laboratory situations (Shorrocks and Swingland, 1990). To mention just a few of the many possible examples: isolated mountain tops, unconnected or poorly connected ponds and lakes, forest patches, human settlements, bird colonies, dung patches and hosts of infectious diseases can all be viewed as habitat islands for certain kinds of organisms – in this sense they are variations of the same theme. The common feature defining such situations as patchy environment problems is that the part of topographical space which offers all necessary resources for population growth is a set of distinct areas (patches) scattered within a continuous matrix of uninhabitable space. In the absence of one or more essential factors of survival or offspring production, individuals cannot survive or multiply within the matrix, but migrants or dispersing propagules can penetrate it. Migrants departing from a habitat patch might survive the journey within the matrix, land on another patch and possibly have offspring there.

Historically, one of the most influential works initiating theoretical research on the subject, and ultimately leading to the present wealth of patch-model approaches in ecology, was obviously a classical series of laboratory experiments due to Huffaker (1958), with a herbivorous mite species feeding on oranges as habitat patches and hunted by a predatory mite. Huffaker expected persistent abundance cycles to emerge in both space and time, but found that it was possible to produce a few complete periods before both mite species died out only if the number of oranges was large enough, and if dispersal among them was constrained by artificial barriers for the predator and facilitated by artificial 'corridors' for the herbivore. Although this experiment did not in fact deliver a sound empirical reinforcement to the intuitive expectation – probably because a large number of relevant effects were unknown or could not be controlled – it did motivate a large body of theory on patch models of predatory systems and patch dynamics in general, which found its way into practical applications as well. Pimentel *et al.* (1963) conducted a similar experiment with flies and their parasites in an artificial system of 'fly-boxes', the connectance of which could be manipulated. The result was similar to that of Huffaker, except that the system was persistent for a longer time. A few of the classic publications on the theory of patch modelling are Reddingius and den Boer (1970), Levin (1974, 1976, 1978), Steele (1974), Hilborn (1975), Hamilton and May (1977), Whittaker and Levin (1977), Crowley (1977, 1978, 1981), Yodzis (1978), DeAngelis *et al.* (1979) and Chesson (1981, 1985).

From a theoretical point of view, perhaps the most important innovation brought forward by patchy environment models is **splitting the spatial scale**: the rules governing the dynamics of a system on the local (within-patch) and the regional (interpatch, or within-matrix) scale are different. Local abundance dynamics can be represented by one of the classical i-state distribution models (hence the name patch-abundance models, as opposed to the spatially implicit patch-occupancy model class), whereas regional dynamics specifies the within-matrix behaviour of the species. The number of different sensible and consistent combinations of local and regional dynamical rules can be large, and the corresponding systems can exhibit a rather colourful range of behaviour, as the models reviewed in this chapter will hopefully illustrate. In fact it is just this wealth of possible combinations that makes patch-abundance approaches useful in modelling rather different ecological problems. The questions such models are designed to answer are the usual problems of population and community dynamics: under what conditions do the populations in question persist or coexist, and what are the stability properties of the model chosen?

The mathematical formulation of patch-abundance problems is relatively straightforward in most cases – in fact this is not the only reason why modellers tend to choose patch modelling when it is one of several possible approaches to the problem at hand. Systems of ordinary differential or difference equations are by far the most commonly applied formal implementations of patch-abundance models. These are usually easy to write down, but their mathematical manipulation and discussion is often anything but straightforward or simple. The 'prototype' systems are usually innocent enough to be fully tractable by paper-and-pencil methods, but some biologically more-inclusive extensions might require the use of computer-orientated numerical techniques in addition to, or as an inevitable substitute for, analytical methods to yield sensible results. This usually turns out to be necessary with the inclusion of more species, more patches, more realistic patterns of population interaction, time delays, refuge effects and the like.

From a practical viewpoint, patch modelling can (and, in fact, it does) contribute to nature conservation issues. Species find themselves in spatially heterogeneous environments in untouched natural habitats, but habitat patchiness is even more pronounced in areas disturbed by, for example, agricultural or industrial activity. These alter the landscape drastically, creating a mosaic of patches that are supportive or hostile, penetrable or deadly for most of the organisms within. The inner boundaries of such artificial mosaic landscapes are very sharp compared to those of natural landscapes, cutting previously continuous habitats into dynamically quite isolated fragments. Habitat fragmentation and its dynamical consequences are a primary concern of recent conservation and landscape ecology (e.g., Fohrig and Merriam, 1985; Hanson et al., 1990), the management proposals of which are in a substantial part based on theoretical insight gained from patchy environment modelling.

3.2 THE PATCH-ABUNDANCE APPROACH

Given the task of incorporating spatial aspects into population dynamical models, it is again a reasonable option to build the representation on classical approaches – that is, to construct models assuming a patchy habitat, but to use well-known classical, non-spatial models to represent the dynamics within the patches. Then it is possible to get a direct insight on the dynamical effects of the spatial structure specified by simply comparing the predictions of the classical model to those of its spatial counterpart. This was the aim of many authors with the study of spatially fragmented populations and communities (e.g., Cohen, 1970; Comins and Blatt, 1974; Roff, 1974; Levin, 1974, 1976, 1978, 1985a; Hamilton and May, 1977; DeAngelis et al., 1979; Hastings, 1982, 1991a, 1993; Crowley, 1981; Comins and Noble, 1985; Diekmann et al., 1988; De Roos et al., 1991; Nisbet et al., 1992; McLaughlin and Roughgarden, 1992).

3.2.1 Basic assumptions of patch-abundance models

The idea behind the patch-abundance approach is that the habitat of a fragmented population or a set of interacting populations consists of distinct patches or islands; these local habitats are dynamically coupled through a certain level of interpatch dispersion of individuals. The fragment of a regional population living on an island is a local subpopulation of the species. The internal dynamics of each subpopulation is represented by one of the popular non-spatial i-state distribution systems such as the Pearl–Verhulst (logistic) or the Gompertz model for single species, or a Lotka–Volterra type system for interacting species' populations. This means that spatial aspects play no role in the autonomous dynamics of a certain patch. It will be also supposed throughout this chapter that local populations are large enough for the omission of stochastic effects due to drift in abundances to be legitimate, and thus a deterministic model to be an acceptable representation of local dynamics. The subpopulations are connected through propagule dispersion among patches, which decreases the net propagule recruitment of the donor subpopulation, but it might increase that of the others. Each patch is a potential propagule acceptor and a potential propagule donor to all others, regarding all species in the system.

Some authors also treat the dynamics within the matrix explicitly, setting up separate equations for within-matrix dynamics besides those of within-patch processes. This is a sensible and in fact an inevitable extension in the case of many invertebrate marine organisms, who spend a substantial part of their lifetime as freely moving pelagic larvae before reaching the sessile adult phase. The equations for within-matrix dynamics represent events of this larval stage. Such specific systems will not be discussed in more detail here; the reader is referred to Roughgarden and Iwasa (1986) and references therein for representative models of this kind.

3.2.2 The general model

Turning first to the simple continuous-time case for a single species on an n-patch habitat, a quite general formulation might be

$$\frac{dx_i}{dt} = F_i(x_i) - e_i(x_i) + \sum_{j \neq i} h_{ij}(x_1, \ldots, x_n), \quad (i \in \{1, \ldots, n\}) \tag{3.1}$$

where x_i (the state variable of the system) is the abundance of the subpopulation resident on patch i. $F(x_i)$ represents the autonomous dynamics that the subpopulation would follow in a closed habitat, and $e_i(x_i)$ and $h_{ij}(x_1, \ldots, x_n)$ are the emigration and immigration terms, respectively. This system is readily extended to an s-species n-patch situation, in which case the (3.1) vector of ordinary differential equations becomes an $n \cdot s$ matrix. The elements of that matrix define the temporal dynamics of the subpopulation of species k resident on patch i:

$$\frac{dx_{ik}}{dt} = F_{ik}(x_{i1}, \ldots, x_{is}) - e_{ik}(x_{i1}, \ldots, x_{is}) + \sum_{j \neq i} h_{ijk}(x_{11}, \ldots, x_{ns}),$$
$$(i, j \in \{1, \ldots, n\}, k \in \{1, \ldots, s\}), \tag{3.2}$$

where $F_{ik}(x_{i1}, \ldots, x_{is})$ is the autonomous growth term for the subpopulation of species k on patch i in the presence of possible interacting subpopulations; $e_{ik}(x_{i1}, \ldots, x_{is})$ and $h_{ijk}(x_{11}, \ldots, x_{ns})$ are the corresponding migration terms. For the most general form (3.2), it is assumed that emigration of species k from patch i depends on the local abundances of all the species on that patch, whereas immigration is a function of all local densities of all species in the system.

3.2.3 The problem of state variable choice

Before going into the details of some specific forms of system (3.2), it is necessary to stop at this point for a moment and take a critical look at the state variable x_{ik}, in relation to the background assumptions necessarily involved in state variable choice. In a non-spatial model, the state variable can be interpreted in two different ways: either as an extensive mass measure (e.g., number of individuals or biomass within habitat) or as an intensive mass-density or simply density measure (like the number of individuals **per unit area of habitat**). Although these two interpretations are equivalent in non-spatial systems, applying a mass-density type state variable is more convenient to use in general, because it allows the modeller not to worry about habitat size at all: the state variable, and, correspondingly, all the extensive parameters of the model are normalized to a unit area of the possibly infinite habitat. In actual fact, the habitat can be assumed to be of any size; this will not affect the formalism because of the overall homogeneity postulated for both the environment and the spatial distribution of the populations.

Such alternative interpretations are more complicated for patch-abundance models like (3.2), if the patches may differ in size. If the state variables of a multipatch model are masses, the parameters in the non-spatial term $F_{ik}(x_{i1}, \ldots, x_{is})$ should be scaled accordingly; for example, the local carrying capacities for a certain species on different patches must be scaled in proportion to patch size (assuming that the patches differ only in size). In this case, the migration parameters can be equal for the same species throughout the patchwork habitat, because the abundance of migrants is also given as a mass measure. If, on the other hand, the state variable is population density, the carrying capacities become intensive parameters; thus they are equal for the same species on different patches, but all the migration terms must be scaled according to patch size. Of course, if the patches are identical in every respect, including size, then the models set up in terms of population masses and population densities do not differ.

3.3 COMPETITION AND MUTUALISM IN DISPERSING ISLAND POPULATIONS

Although the classical Lotka–Volterra model of two-species competition is a standard subject of any basic textbook on mathematical ecology or population biology (cf., for example, May (1976) and Begon *et al.* (1986)), it is worth while to recall its main conclusions, together with some more recent and more general results on non-spatial competition–mutualism models of

Population dynamics in patchy environments

the Lotka–Volterra type, because these will serve as a ground for comparison with the results of competitive patch-abundance models.

The non-spatial two-species model is given by the pair of ordinary differential equations

$$\frac{dx_1}{dt} = r_1 x_1 \left(1 - \frac{x_1 + \alpha_{12} x_2}{K_1}\right)$$
$$\frac{dx_2}{dt} = r_2 x_2 \left(1 - \frac{x_2 + \alpha_{21} x_1}{K_2}\right),$$
(3.3)

in which x_k are the densities of the competing populations, with growth constants $r_k > 0$ and carrying capacities $K_k > 0$. $\alpha_{kl} > 0$ are the competition parameters, which can be interpreted as the density of species l that is equivalent – in terms of resource depletion, for example – to a unit density of species k. Substituting $g_{kk} = \frac{r_k}{K_k}$ and $g_{kl} = \frac{r_k \alpha_{kl}}{K_k}$, $k \neq l$, into (3.3) yields the equivalent form

$$\frac{dx_1}{dt} = x_1(r_1 - g_{11} x_1 - g_{12} x_2)$$
$$\frac{dx_1}{dt} = x_2(r_2 - g_{21} x_1 - g_{22} x_2).$$
(3.4)

This system has always three equilibria on the boundary of the positive quadrant, namely $(0, 0)$; $(K_1, 0)$; $(0, K_2)$. It admits also an interior (that is, positive) equilibrium point $(\hat{x}_1, \hat{x}_2)$, if the zero-growth isoclines defined by

$$r_1 - g_{11} x_1 - g_{12} x_2 = 0 \quad \text{for species 1}$$
$$r_2 - g_{21} x_1 - g_{22} x_2 = 0 \quad \text{for species 2}$$
(3.5)

intersect inside the positive quadrant; that is, if the solution of the two-dimensional linear equation (3.5) is positive. Using Cramer's rule, this translates to

$$\frac{r_1 g_{22} - r_2 g_{12}}{g_{11} g_{22} - g_{12} g_{21}} = \hat{x}_1 > 0 \quad \text{and} \quad \frac{r_2 g_{11} - r_1 g_{21}}{g_{11} g_{22} - g_{12} g_{21}} = \hat{x}_2 > 0.$$
(3.6)

The origin (trivial equilibrium) is always unstable; the interior fixed point, $(\hat{x}_1, \hat{x}_2)$, if it exists, can be either stable or unstable. The simple necessary and sufficient condition for its stability, that is, for the coexistence of the two competitors, is

$$\frac{r_1}{g_{12}} > \frac{r_2}{g_{22}} \quad \text{and} \quad \frac{r_2}{g_{21}} > \frac{r_1}{g_{11}}.$$
(3.7)

In this case, both boundary equilibria $(K_1, 0)$ and $(0, K_2)$ are unstable, that is, the single-species fixed points are invadable. It can be shown (with the use of Liapunov functions) that if $(\hat{x}_1, \hat{x}_2)$ is locally stable, it is also globally stable; that is, it is approached by all the orbits starting from within the positive quadrant of the state space.

If $(\hat{x}_1, \hat{x}_2)$ exists, but is unstable (saddle point), the outcome of competition depends on the initial densities: the system approaches either $(K_1, 0)$ or $(0, K_2)$, both stable in this case (bistability); their basins of attraction are separated by the smooth curve connecting the origin and $(\hat{x}_1, \hat{x}_2)$. The separatrix contains the only two orbits which approach asymptotically the – otherwise unstable – fixed point.

The isoclines do not cross within the positive quadrant of the state space if (3.6) does not hold. Then one of the two non-trivial boundary equilibria $(K_1, 0)$ or $(0, K_2)$ is stable, that is, one species excludes the other. Which of them wins the competition depends on which isocline falls further from the origin. Both these outcomes are illustrated on Figure 3.1, together with the coexistent and the bistable cases.

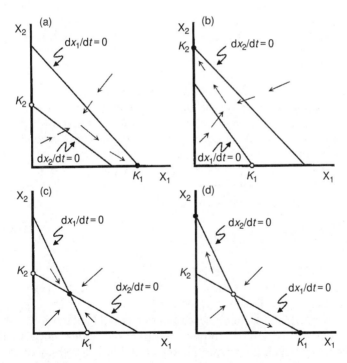

Figure 3.1 Equilibria of the non-spatial two-species Lotka–Volterra competition model: (a) and (b) one of the species wins, the other goes extinct; (c) unconditional coexistence; and (d) conditional extinction (bistable case); which of the competitors wins depends on the initial densities.

The s-species generalization of (3.4),

$$\frac{dx_k}{dt} = F_k(x_1, \ldots, x_s) = x_k\left(r_k - \sum_{l=1}^{s} g_{kl} x_l\right) \quad k = (1, \ldots, s) \qquad (3.8)$$

keeps the property of admitting one positive fixed point at most, but also a great number ($2^s - 2$) of non-trivial boundary equilibria are possible with one or more species extinct, others present. The necessary and sufficient condition for the existence of a positive equilibrium is that there be a positive solution $\hat{\mathbf{x}} > \mathbf{0}$ to the s-dimensional linear equation

$$r_k - \sum_{l=1}^{s} g_{kl} \hat{x}_l = 0, \quad k = 1, \ldots, s, \qquad (3.9)$$

which appears in matrix algebraic notation as

$$\mathbf{r} - \mathbf{G}\hat{\mathbf{x}} = \mathbf{0}. \qquad (3.10)$$

$\mathbf{G}$ is called the community matrix of system (3.8), containing the constant coefficients of pairwise population interactions g_{kl} as its elements. Thus the general condition for the existence of the persistent equilibrium is

$$\hat{\mathbf{x}} = \mathbf{G}^{-1}\mathbf{r} > \mathbf{0}. \qquad (3.11)$$

Equation (3.6) is a special case of (3.11) with $s = 2$.

The local stability of $\hat{\mathbf{x}}$ (or of any other fixed point, if it exists) can be checked using the Jacobian matrix $\mathbf{A}$ (Appendix B) of the function $\mathbf{F}$, evaluated at $\hat{\mathbf{x}}$ (or at the corresponding fixed point). Note that in the specific case of the Lotka–Volterra competition model, the community matrix $\mathbf{G}$ and the Jacobian matrix $\mathbf{A}$ are connected through the equality $\mathbf{A}(\hat{\mathbf{x}}) = -\hat{\mathbf{x}} \cdot \mathbf{G}^{-1}$. $\hat{\mathbf{x}}$ is locally stable if the dominant (of largest real part) eigenvalue of $\mathbf{A}(\hat{\mathbf{x}})$ has a negative real part:

$$\mathrm{Re}\left[\lambda_{\max}\left(\mathbf{A}(\hat{\mathbf{x}})\right)\right] < 0 \qquad (3.12)$$

Applying (3.12) to the two-species system (3.4) leads to (3.7), the condition for the stability of the positive two-species fixed point. However, for $s > 2$, (3.12) no longer guarantees the global stability of the interior fixed point.

Little is known about the dynamics of non-spatial competitive systems in general, and also of Lotka–Volterra competition models in particular, for more than two species. The central result (Hirsch, 1988) comes from the study of the competitive system

$$\frac{dx_k}{dt} = x_k \cdot f_k(x_1, \ldots, x_s), \quad k = 1, \ldots, s, \qquad (3.13)$$

in which f_k, the actual **per capita** growth rate of species k, is a general form for competitive interaction – not necessarily of the Lotka–Volterra type, linear function. All that is assumed for f_k is that

1. it decreases monotonically in each dimension of the state space: $\partial f_k/\partial x_l < 0$ for any competitor l (including k itself),
2. it must become negative for large x_k even if no other competitors are present, and
3. the origin is repelling into the positive orthant, that is, $f_k(\mathbf{0}) > 0$ for all k.

Hirsch (1988) shows that, under these assumptions, the state space contains an $(s-1)$-dimensional surface attracting all non-negative orbits, except for the one starting (and staying) in the origin. All possible limit sets (fixed points, limit cycles, strange attractors, etc.) of the s-dimensional competitive system lie on this hypersurface. The attracting hypersurface can be viewed as the 'joint carrying capacity' or 'carrying simplex' of the s-dimensional competitive system (3.13). It is in fact analogous to the carrying capacity of a single-species logistic model in the sense that orbits starting either under or above the carrying simplex end up on it asymptotically. The picture is similar also in higher dimensions: the attracting hypersurface is spanned by the single-species carrying capacities (single-species equilibria) on the axes of the state space.

Smale (1976) published a complementary result earlier, proving that for **any** bounded dynamical system confined to the $(s-1)$-dimensional simplex, there exists an s-dimensional competitive system of the (3.13) type, which has the original $(s-1)$-dimensional dynamics on its carrying simplex as its limit set. This amounts to the surprising statement that – in principle at least – a non-spatial ordinary differential equation (ODE) model of a bounded system of s-species ecological interactions can always be reformulated as the limit set of an $(s+1)$ species competition model. To find the corresponding $(s+1)$ species competition model is, of course, a different – not nearly trivial – problem.

The theorem of Hirsch (1988) confines the limit sets of s-dimensional competitive systems onto the $(s-1)$-dimensional hypersurface, but it does not say anything about the behaviour of the orbits **within** the carrying simplex. The most general statement on this is a corollary of the Hirsch (1988) result and the Poincaré–Bendixson theorem (cf., for example, Hirsch and Smale (1974)). The latter states that **whatever** dynamics we consider on a two-dimensional surface, the orbits either come arbitrarily close to some fixed point, or they are closed – no open limit sets such as chaotic attractors can exist in two dimensions. Since the competition model (3.13) with three species can be reduced to a two-dimensional system by considering the carrying simplex as its state space (which the orbits do approach anyway), the Poincaré–Bendixson theorem guarantees that a three-species competitive system can admit limit cycles at most: no strange attractors can arise. Hofbauer and So (1994) give a three-species Lotka–Volterra example (3.8), which shows at least two persistent limit cycles around the stable positive fixed point, the inner one repelling, the outer attracting. The Poincaré–Bendixson theorem does not apply, however, to competition models with $s > 3$: persistent chaotic orbits have been proven to exist even in a simple four-dimensional case of (3.8) (Arneodo et al., 1982).

3.3.1 The multispecies multipatch Lotka–Volterra model

Levin (1974), Slatkin (1974), DeAngelis et al. (1979) and others consider a specific form of (3.2), coupling Lotka–Volterra type competitive–mutualistic local systems with simple linear immigration terms. DeAngelis et al. (1979), in particular, use the following form:

$$\frac{dx_{ik}}{dt} = x_{ik}\left(r'_{ik} - \sum_{l=1}^{s} g_{ikl} x_{il}\right) + \sum_{j=1}^{n} h_{ijk} x_{jk}. \tag{3.14}$$

Implicit in the population growth parameter r'_{ik} of species k on patch i is the emigration parameter e_{ik}, which is assumed to be the constant of the linear emigration term $-e_{ik}x_{ik}$, so that

$$r'_{ik} = r_{ik} - e_{ik}, \tag{3.15}$$

where r_{ik} is the intrinsic rate of population growth for species k on patch i. g_{ikl} is the coefficient of the local biotic effect (competitive or mutualistic) exerted by species l on species k, on patch i. Thus the first term in (3.14) represents the autonomous dynamics of species k on patch i, including population growth due to natality, abundance loss due to mortality and emigration to other patches, and local biotic interactions. The second term is the contribution of all other patches to the population growth of species k on patch i through immigration. Substituting (3.15) into (3.14), the latter can be rewritten with all three terms of population growth, emigration and immigration separated as

$$\frac{dx_{ik}}{dt} = x_{ik}\left(r_{ik} - \sum_{l=1}^{s} g_{ikl} x_{il}\right) - e_{ik}x_{ik} + \sum_{j=1}^{n} h_{ijk} x_{jk}. \tag{3.16}$$

The constants e_{ik} and h_{ijk} are all non-negative. Considering this extended form, equivalent to system (3.14), it is easy to see that the emigration and immigration parameters e_i and h_{ji} are constrained so that the outflow of species k by emigration from a certain island i must be greater than or equal to the sum of inflows from that island to all others:

$$e_{ik} \geq \sum_{j=1}^{n} h_{ijk}, \quad (i \neq j, \, i = 1, \ldots, n). \tag{3.17}$$

The strict > relation implies that a fraction of the emigrants gets lost during migration, whereas the = relation conforms the assumption that all migrants arrive at one of the patches safely. Note that migratory movements are

3.3.2 Persistence and coexistence conditions

DeAngelis et al. (1979) study the local persistence and stability properties of this model, using local stability analysis for (3.14). The system is considered to be persistent if the origin of the state space, the trivial equilibrium point, is repelling into the positive orthant so that no species die out completely from the patchwork. In other words, persistent trajectories move off the state of global extinction, however close to that they were driven by external forces. This applies to the whole species assemblage: for each species $k \in \{1, \ldots, s\}$, there exists at least one patch $i \in \{1, \ldots, n\}$ such that

$$\liminf_{t \to \infty} x_{ik}(t) > 0.$$

Stability analysis by local linearization is a simple matter for systems of low dimensionality (see Appendix B). Most linearized systems become intractable analytically very soon, however, as the number of dimensions (here, the number of species and patches involved) increases. Local stability analysis of (3.14) requires the eigenanalysis of the $(ns \times ns)$ Jacobian matrix $\mathbf{A}(E)$, evaluated at each possible equilibrium point E of the system. All stability tests are related to the sign of the dominant eigenvalue of the Jacobian, which is in turn related to the sign structure of the Jacobian itself. The problem in hand is to find practical stability conditions for (3.14), in terms of the sign structure of $\mathbf{A}(E)$, that are relatively easy to check even if $\mathbf{A}$ is large.

For the special case of single-species multipatch situations ($s = 1$), (3.14) simplifies to

$$\frac{dx_i}{dt} = x_i(r_i' - g_i x_i) + \sum_{j=1}^{n} h_{ij} x_j. \tag{3.18}$$

Freedman and Takeuchi (1989a) and Hofbauer (1990) have proven that if (3.18) or a similar monotone single-species system is persistent, then there is a fixed point within the positive orthant which is both unique and stable, that is, the single positive equilibrium of the system is globally stable (cf. also Lu and Takeuchi, 1993).

Thus the only condition for (3.18) to be globally stable in the positive orthant is that it be persistent – that is, a proof of persistence is also the proof of global stability. Persistence is easy to prove by the local stability analysis of (3.18) at the origin: if the origin is locally unstable, the system is persistent (persistence meaning that the species is present on at least one patch). The Jacobian matrix of (3.18) is of a special sign pattern, all diagonal elements being negative, and all off-diagonal elements non-negative. This simplifies the stability analysis of the origin (and also that of any other fixed point) a great deal: it is sufficient to determine the sign structure of the principal minors for the Jacobian at the origin (Box 3.1) to establish persistence conditions. This method will be applied below to a specifically constrained form of the multispecies extension of (3.18).

For the multispecies model (3.14) in a purely competitive setting (with s species competing on n habitat patches, no mutualisms; $g_{ikl} \geq 0$), the problem of persistence can be tackled on a species-by-species basis. Suppose first that the species do not compete at all ($g_{ikl} = 0$ for all index triplets ikl), that is, they can coexist on the n patches without any interaction. Then (according to the global stability result above) if species k is persistent, the corresponding single-species multipatch system (3.18) approaches a positive fixed point $\tilde{\mathbf{x}}_k = (\tilde{x}_{1k}, \ldots, \tilde{x}_{nk})$. If k is not persistent in the absence of competitors, it will be even less so under competitive pressure: species k dies out also from (3.14) in this case. Suppose that all s species are persistent without competition; $\tilde{x}_{ik}$ is the non-competitive local equilibrium of species k on patch i. Now let us return to the competitive case with $g_{ikl} \geq 0$. The worst possible biotic environment that a rare (close to zero abundance) local population of species k on patch i can experience is $\tilde{E}_{ik} = (\tilde{x}_{i1}, \ldots, \tilde{x}_{i,k-1}, 0, \tilde{x}_{i,k+1}, \ldots, \tilde{x}_{is})$, because each local population can safely be assumed to stay below its non-competitive equilibrium abundance under competitive pressure. Now substituting $\tilde{E}_{ik}$ into the competition term of (3.14) on all patches yields

$$\frac{dx_{ik}}{dt} = x_{ik}\left(r'_{ik} - \sum_{\substack{l=1 \\ l \neq k}}^{s} g_{ikl}\tilde{x}_{il}\right) + \sum_{j=1}^{n} h_{ijk}x_{jk}, \quad (i = 1, \ldots, n) \tag{3.19}$$

which is the model of the dynamics of species k within the patchwork, assuming the strongest possible competitive pressure it might face. The intuitive expectation that a local population which is able to grow under such pessimal biotic conditions will perform better in any actual situation proves to be right in this case: DeAngelis et al. (1979) show that a sufficient condition for the simultaneous persistence of all s species within the patchwork is that (3.19) be persistent for all k (see Box 3.2). In fact they prove an even stronger statement: this sufficient condition is also necessary for (3.14) to be persistent.

For the general multispecies–multipatch competitive–mutualistic situation, practically useful criteria for checking the stability of positive equilibria are known only for community matrices representing a so-called 'limited competition' case (DeAngelis et al., 1979; Travis and Post, 1979; Smith, 1986, 1988) in which the sign structure of the community matrix is constrained in a specific manner. An explicit definition for limited competition in a multipatch system is that

1. the signs of interaction coefficients between any pair of species be symmetric on a certain patch, and, for a certain species pair, be the same on all patches, that is,

Competition and mutualism

Box 3.1

For (3.18), the sign structure of the Jacobian matrix $\mathbf{A}(E)$ takes a special form for all possible equilibrium points $E = (\hat{x}_1, \ldots, \hat{x}_n)$: the off-diagonal elements h_{ij} of $\mathbf{A}(E)$ are all non-negative:

$$A(E) = \left[\frac{\partial \dot{x}_i}{\partial x_j}\right]_{\hat{x}_i} = \begin{bmatrix} r_1' - 2g_1\hat{x}_1 & h_{12} & \cdots & h_{1n} \\ h_{21} & r_2' - 2g_2\hat{x}_2 & \cdots & h_{2n} \\ \vdots & \vdots & \ddots & \vdots \\ h_{n1} & h_{n2} & \cdots & r_n' - 2g_n\hat{x}_n \end{bmatrix}. \quad [3.1]$$

For a Jacobian matrix of non-negative off-diagonal elements, a local condition for the stability of E can be formulated in terms of the leading principal minors (Appendix C) of $\mathbf{A}(E)$ (DeAngelis et al., 1979; Smith, 1988). The simple necessary and sufficient condition for the linearized system to be locally stable around an equilibrium point E is that the leading principal minors of $\mathbf{A}(E)$ alternate in sign, that is:

$$(-1)^i \det \begin{bmatrix} r_1' - 2g_1\hat{x}_1 & h_{12} & \cdots & h_{1i} \\ h_{21} & r_2' - 2g_2\hat{x}_2 & \cdots & h_{2i} \\ \vdots & \vdots & \ddots & \vdots \\ h_{i1} & h_{i2} & \cdots & r_i' - 2g_i\hat{x}_i \end{bmatrix} > 0, \quad (i = 1, \ldots, n)$$

[3.2]

When checking the persistence of the system, E represents the origin of the state space (trivial equilibrium), that is, $\hat{x}_i = 0$ for all $i \in \{1, \ldots, n\}$, and the diagonal elements of the corresponding Jacobian matrix $\mathbf{A}(0)$ are r_i'; the terms for density dependence cancel. Note that if any one of the values of r_i' is positive then the system is persistent on at least the corresponding patch, even without immigration input from other patches, and thus it is also persistent within the whole patchwork. Also with all r_i' parameters negative, if any one of the leading principal minors of $\mathbf{A}(0)$ violates [3.2], the origin is a repeller, meaning that the species is persistent. For checking the local stability of a positive equilibrium point $E = (\hat{x}_1, \ldots, \hat{x}_n)$, the leading principal minors have to be evaluated for the Jacobian at E, but the method is exactly the same otherwise.

> **Box 3.2**
>
> A necessary and sufficient condition for species k to become extinct (the origin to be an attractor) within the patchwork is similar to [3.2]:
>
> $$(-1)^i \det \begin{bmatrix} r'_{1k,\text{eff}} & h_{12k} & \cdots & h_{1ik} \\ h_{21k} & r'_{2k,\text{eff}} & \cdots & h_{2ik} \\ \vdots & \vdots & \ddots & \vdots \\ h_{i1k} & h_{i2k} & \cdots & r'_{ik,\text{eff}} \end{bmatrix} > 0, \quad (i=1,\ldots,n), \qquad [3.3]$$
>
> where
>
> $$r'_{jk,\text{eff}} = r'_{jk} - \sum_{\substack{l=1 \\ l \neq k}}^{s} g_{jkl}\tilde{x}_{jl}, \quad (j=1,\ldots,i) \qquad [3.4]$$
>
> is the effective growth rate of species k on patch j in the presence of all competing species l at their autonomous local equilibrium abundances $\tilde{x}_{jl}$, that is, in a pessimal biotic environment. If any one of criteria [3.3] is violated, species k is persistent.

$\text{sign}(g_{ikl}) = \text{sign}(g_{ilk})$ and $\text{sign}(g_{ikl}) = \text{sign}(g_{jkl})$

for all $i,j \in \{1,\ldots,n\}$, $k,l \in \{1,\ldots,s\}$, and

2. within-patch sign structure of biotic interactions be such that 'friends of friends and enemies of enemies be friends, and friends of enemies and enemies of friends be enemies' for each species, that is,

$\text{sign}(g_{ikl}) = \text{sign}(g_{ikm})\text{sign}(g_{iml})$

for all $k,m,l \in \{1,\ldots,s\}, k \neq m \neq l, i \in \{1,\ldots,n\}$.

Note that condition 1 automatically excludes predative and parasitic interactions, these being asymmetric. Limited competition emerges, for example, in a non-spatial system divided into two species groups, within each group only mutualistic (+, +) or neutral (0, 0) interactions taking place, and competition (−, −) occurring only between species that belong to different groups. In general, if the limited competition conditions 1 and 2 hold for the interaction coefficients g_{ikl} of a system (3.14), then the local stability of **any** equilibrium point of the corresponding system can be checked by applying simple leading principal minor criteria analogous to [3.2] (Box 3.1) (for more details, see DeAngelis *et al.*, 1979 and references therein).

3.3.3 Single species persistence: source–sink dynamics

DeAngelis et al. (1979) demonstrate for the single species n-patch model that even if the local subpopulations themselves go extinct if left alone (that is, all values of r'_i are negative), a subset of the habitat islands can guarantee persistence for the species, given sufficient flow of abundance among them (large enough h_{ij} parameters). The assumption of a large propagule input means an intensive exchange of abundance among persistent subpopulations, since e_i must also be large for some of h_{ji} to be large (cf. (3.17)).

The only non-trivial case of interest for discussion is that with negative effective growth rates on all patches, that is, $r'_i = r_i - e_i < 0$ for all i. One would expect by intuition, however, that at least one subpopulation should have a positive intrinsic growth rate r_i to ensure the persistence of the species within the patchwork. Habitat patches with $r_i > 0$ represent 'source habitats', and those with $r_i < 0$ are 'sinks' *sensu* (Pulliam, 1988). The presence of at least one source is necessary, but not sufficient in itself for the species to persist; if a source dissipates too many emigrants to patches which are unable to maintain a persistent subpopulation in spite of the immigration input, the source subpopulation will itself also go extinct. A source subpopulation must have other sources, or at least not very 'absorptive' sinks around to persist, from which it can gain immigrants to increase its effective local growth rate above zero. In other words, sources need 'mirror' patches that reflect back a fraction of the emigrants they emit, whereas persistent sinks are in need of a steady net abundance gain from immigration to compensate for their autochthonous net loss of abundance.

3.3.4 Habitat fragmentation effects

The effect of habitat fragmentation on the propensity for regional persistence of a single species can be considered in relation to the (3.18) model, by making plausible assumptions regarding changes in model parameters as a result of fragmentation. From a topographical point of view, habitat fragmentation implies patch size reduction and interpatch distance increases simultaneously. Patch size reduction decreases the carrying capacity r_i/g_i of the patch (assuming the state variable to be of the mass type), which translates to an increase in g_i. This alone does not affect persistence directly, since conditions [3.3] (Box 3.2) are independent of g_i at the trivial fixed point $E = (0, \ldots, 0)$. Larger g_i values decrease all local abundances at any positive equilibrium (if there is such), however, which makes the system less resistant to external disturbances and demographic stochasticity (May, 1976).

It is reasonable to assume that decreasing the size of the habitat will at least not improve the performance of a subpopulation in terms of increasing its intrinsic growth rate r_i, provided that the environment does not change otherwise. The parameters of migration, e_i and h_{ji}, might be strongly affected, however, which is very easy to see, for example, if the individuals move at random both within and among the patches. Emigration is likely to increase in this case, since the perimeter-to-area ratio of a smaller patch is higher; therefore, a greater proportion of the subpopulation will encounter and cross the boundary. Randomly moving immigrants are obviously less likely to land on smaller habitats. Both these effects on migration are adverse as regards the persistence of a population in a fragmented habitat, since they reduce the effective growth rate of the subpopulation, and decrease the connectance of the patches.

Interpatch distance can be safely assumed not to affect within-patch parameters, only those connecting habitats, that is, values of h_{ji}. If the assumption of a fast random movement of the individuals within the region holds, even these can be postulated to be distance-independent, as is the case also with metapopulation models (cf. Chapter 4). The alternative, and more realistic, assumption of a decreasing distance dependence function for h_{ji} again suggests that habitat fragmentation might be a threat of extinction. This is a conclusion also suggested on the basis of quite a few field studies (Pickett and Thompson, 1978; Quinn and Hastings, 1987; Quinn and Harrison, 1988; Burkey, 1989; Quinn *et al.*, 1989; Doak *et al.*, 1993).

3.3.5 Spatial pattern and competitive coexistence

For the simultaneous persistence of competing species, that is, for competitive coexistence to occur, conditions [3.3] (Box 3.2) should be transgressed by more than one species at the same time. A simple parameter combination for species k violating [3.3] is one with not-too-large negative effective local growth rates $r'_{i,k,\text{eff}}$, and intensive dispersion (h_{ijk} large). Effective growth rates take moderate values if the competition parameters g_{ikl} are small. In all, weak interspecific competition and sufficient migration among habitat patches favours the coexistence of competing species in a patchy environment. Comparing this conclusion to those of the non-spatial Lotka–Volterra competition model (cf. Section 3.2.1; also Strobeck (1973)) and the single-species patch model above, this result is not surprising. There are some specific cases related to different spatial abundance configurations, however, which might be of more interest from an ecological point of view.

First consider the case without interspecific competition (e.g., suppose that the species utilize different resources throughout the patchwork). Then the dynamics of the species are disconnected; each can be studied independently from all others, using the single-species multipatch model (3.18). If all species are persistent within the patchwork, they coexist on the 'regional' scale of the whole set of patches. This does not, however, necessarily mean that they really coexist also on the 'local' scale of single patches: one species might be dominant on a given subset of habitats and 'accidental' (very rare) on the rest. The abundance patterns of the different species do not even need to overlap. In the special situation of each species dominating its own habitat subset, one might guess that introducing reasonable measures of interspecific competitive interactions will not significantly affect the outcome: starting from a point sufficiently close to the joint equilibrium of the non-competitive system, trajectories might tend to the same equilibrium. In such a case, species would coexist regionally by means of local segregation. In the general case with each habitat potentially supporting 0 to s species in considerable abundance at the non-competitive equilibrium, the larger the habitat overlap between two species, the smaller we might expect to be the domain of parameter space that permits their simultaneous regional persistence (regional coexistence) in a competitive situation. This suggestion awaits formal analysis, however. What effect on persistence and coexistence we can get by increasing the dispersal connectance of patches in this case is an open question as well. The answer might be highly dependent on what specific system is actually analysed.

Even though persistent single-species systems themselves admit unique, globally stable positive equilibria in non-competitive situations, it is possible that multiple fixed points (or, in some instances, also more complex limit sets, e.g., limit cycles or strange attractors) emerge with competition introduced, so that the state space of the joint system consists of multiple basins of attraction. Initial density pattern may be crucial in determining which fixed point (or limit set) a certain trajectory approaches in such a system.

Taking a different (habitat-orientated) viewpoint instead of the species-based one above, consider the model (3.14) with competition, but without immigration (that is, $h_{ij} = 0$ for all i, j; all emigrants are lost during dispersal). Now the dynamics of the habitat patches are disconnected, and the classical multispecies single patch (that is, non-spatial) Lotka–Volterra model applies independently to each isolated patch. Then, of course, only species with at least one self-maintaining subpopulation can be persistent, that is, those with a positive effective growth rate r'_{eff} on at least one patch. If the patches are identical environmentally, but the corresponding non-spatial multispecies system admits, for example, persistent periodic solutions within the positive orthant, or if the patches differ environmentally, each maintaining its own stable equilibrium community, the resulting abundance pattern within the patchwork might be heterogeneous. Levin (1978), using a more general – not necessarily monotone – model, shows that in a system of such isolated, self-maintaining local communities, the introduction of linear migration terms with small exchange (migration) coefficients does not alter the qualitative behaviour of a stable equilibrium, if there is such. A stable non-migratory heterogeneous pattern of abundance is preserved and only slightly and quantitatively affected, if migration movements among patches are rare: weak-enough dispersal coupling of local communities will not destroy the stable abundance pattern.

3.3.6 Single species resilience and risk spreading

Persistence and stability are concepts that possess a definite meaning both in ecological and in mathematical terms. In contrast, resilience is a more intuitive concept without a rigorous mathematical definition, roughly meaning how fast a system returns to its stable limit set (fixed point, limit cycle etc.) after perturbations. This question is always secondary to the problems of persistence and stability, because – in a purely deterministic model – resilience does not affect the predictions qualitatively. If, however, we allow for stochastic effects when population abundances are low, resilience becomes important even in this respect. How much time a trajectory spends close to the boundary of the state space is an important question in this case, since the boundary might eventually absorb, by chance, trajectories that run nearby for a long time. Reaching the boundary of the state space represents the extinction of one or more species from the model community.

Vance (1984) examines a set of single-species multipatch systems with optionally density-dependent birth, death and dispersal rates. His focus lies on the effect of a range of assumptions regarding dispersal mechanism on population resilience against heterogeneous perturbations. (Heterogeneous perturbation means that perturbations can be different on different

---

Box 3.3

The assumptions of identical patches (identical diagonal elements of $A(N^*)$) and obligate, indiscriminate dispersal (identical off-diagonal elements of $A(N^*)$) assure that there are only two kinds of eigenvalues to $A(N^*)$; the leading eigenvalue, λ_1, is different from the $(n-1)$ other eigenvalues, themselves equal:

$$\lambda_1 = -(b-d)$$
$$\lambda_k = -(2b-d), \quad k = (2,\ldots,n) \tag{3.5}$$

The resilience of the single-patch equivalent of (3.20) (that is, the logistic model) depends only on the leading eigenvalue λ_1 (note that $(b-d)$ is the intrinsic growth rate of the population), whereas the extra resilience against heterogeneous perturbations gained by dispersal as compared to the resilience of the non-dispersive case is related to λ_k ($k \neq 1$). In both cases, the more negative the eigenvalue, the more resilient the system will be, meaning that all the perturbed trajectories will approach the fixed point faster. If the absolute value of λ_k ($k \neq 1$) is much larger than that of λ_1 (that is, if b is large), dispersal adds a large extra resilience to the heterogeneously perturbed system.

patches.) A simple model is discussed below from among those given by Vance (1984).

Consider a habitat consisting of n similar patches, and a single species growing and dispersing among them. Births are density- and patch-independent, and all newborn individuals disperse evenly among the patches without loss during migration. Death rates are density-dependent with the same coefficients on each patch. A continuous time model with these assumptions is

$$\frac{dN_i}{dt} = (1/n)\sum_{j=1}^{n} bN_j - (d + d_1 N_i)N_i, \quad i = (1, \ldots, n) \tag{3.20}$$

where N_i are local abundances, b is the constant birth rate, d and d_1 are death rate constants, each non-negative, and $(b - d) > 0$ to ensure population growth at low densities. With this mechanism, each local population approaches the uniform equilibrium density

$$N^* = (b - d)/d_1 \tag{3.21}$$

in the limit as $t \to \infty$. This fixed point is globally stable, just like that of the logistic equation (which is a special case of (3.20) with $n = 1$).

The sign of the real part of the leading eigenvalue of $\mathbf{A}(N^*)$, the Jacobian matrix (cf. Box 3.1) for (3.20), evaluated at the equilibrium (3.21), deter-

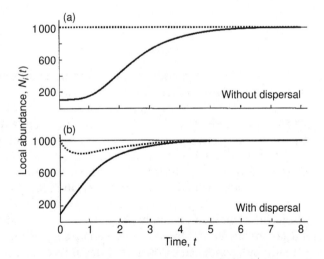

Figure 3.2 The effect of dispersive coupling to an undisturbed population (dotted line) on the resilience of a severely perturbed population (solid line). (a) The recovery process without coupling; and (b) recovery with coupling. Model parameters: $b = 1$, $d = 0$, $d_1 = 0.001$. $N^* = 1000$; eigenvalues: $\lambda_1 = -1$, $\lambda_2 = -2$ (after Vance (1984)).

78   Population dynamics in patchy environments

mines the local stability of N^*. The absolute magnitude of the leading eigenvalue and that of the other eigenvalues at the fixed point can be used as resilience measures for the system (Box 3.3).

Figure 3.2 shows a pair of graphs with numerical solutions of (3.20) for two patches ($n = 2$), comparing the resilience against a heterogeneous perturbation (whereby only one of the local populations is perturbed) of a non-dispersive model to that of the corresponding dispersive system. (Homogeneously perturbed systems behave obviously just like the single-patch logistic model.) Figure 3.2 suggests that the extra resilience against heterogeneous perturbations is a clear manifestation of the risk-spreading effect (den Boer, 1968, 1970), in the sense that the reduced risk of extinction on the perturbed patch shows up at the expense (that is, increased risk of extinction) of the undisturbed population. It is also obvious from [3.5] (Box 3.3) that the difference between the leading eigenvalue and the rest depends on the birth rate constant b alone: the larger it is, the more pronounced will be the risk-spreading effect.

3.4 PREDATION IN PATCHY HABITATS

The origin of almost all theoretical studies on the population dynamics of predation is the classical pair of differential equations

$$\frac{dH}{dt} = rH - cHP$$
$$\frac{dP}{dt} = -dP + fcHP, \qquad (3.22)$$

the non-spatial Lotka–Volterra model for predation describing the abundance dynamics of a predator population feeding exclusively on a single prey species (cf., for example, Nisbet and Gurney, 1982; Hofbauer and Sigmund, 1988; Murray, 1989). In (3.22), H and P are the abundances of the prey and the predator, respectively; r is the constant (density-independent) growth rate of prey in the absence of predators, c is the attack rate of predators on prey, d is the constant death rate of the predator without prey present and f is a constant factor for captured prey abundance conversion to predator abundance. All constants are positive. This classical model of predation is built on a number of simplifying assumptions besides overall environmental homogeneity and the absolute monophagy of the predator, namely the density independence of birth and death rates regarding both species, and an unlimited linear predator functional response to prey density. Density independence implies that the interacting populations represent the only limiting factors one to the other: the prey population is limited by the abundance loss due to predator activity, whereas the predator population can maintain itself only in the presence of prey.

Being non-linear in H and P, the general solution for even this oversimplified system is unknown in algebraic form. Apart from the trivial equilibrium $E = (0, 0)$, which is a saddle point, (3.22) has another fixed point within the positive quadrant of the state space: $\hat{E} = (d/fc, r/c)$. Stability analysis of the system around $\hat{E}$ by local linearization (see Appendix B) indicates that the orbits are periodic and that they might be closed, since the conjugate pair of leading eigenvalues of the Jacobian matrix are purely imaginary. The constant of motion for (3.22) is known (cf. Hofbauer and Sigmund, 1988), proving that the system is conservative, that is, the orbits are in fact closed around the fixed point: (3.22) is neutrally stable. Figure 3.3 is a phase portrait of (3.22).

Neutral stability is a somewhat odd behaviour from an ecological point of view – in fact, no attempts to find relevant examples of it in the field or laboratory have been successful. It is, however, an optimal starting point for assessing the effects on stability of different extensions and modifications to the original system, because it is not robust: the stability type of the unique interior fixed point is very sensitive to structural changes in the model. These might involve, for example, the inclusion of density-dependent population growth, bounded predator functional response to prey density, and, most important from our present aspect, patchy habitats. Before going into

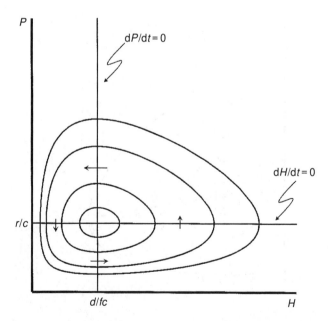

Figure 3.3 Phase portraits of the non-spatial predator–prey Lotka–Volterra model (3.22) with different initial conditions.

80   Population dynamics in patchy environments

the details of the spatially structured version, results of some non-spatial modifications should be discussed here briefly for later reference.

Introducing density-dependent self-regulation into any one or both of (3.22) proves to be stabilizing, that is, these transform the neutrally stable positive fixed point into an asymptotically stable one. The modified version of (3.22) with both predator and prey density dependence assumed is

$$\frac{dH}{dt} = rH\left(1 - \frac{H}{K}\right) - cHP$$
$$\frac{dP}{dt} = -d(1+gP)P + fcHP$$
(3.23)

where $K > 0$ is the constant carrying capacity of the habitat regarding the prey species, and $g \geq 0$ is the slope of the linear density dependence function for predator death rate. A phase portrait for this system is given on Figure 3.4.

If we do not allow for the **per capita** predation rate of the predator to exceed a certain level (which is a reasonable assumption, given the inevitable physical and physiological limitations of capturing, handling and ingestion of prey), the unbounded linear function cH in the predation terms of the previous models should be replaced by some bounded saturation func-

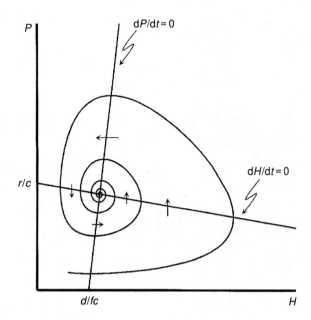

Figure 3.4 Phase portrait of the non-spatial Lotka–Volterra model of predation with prey and predator density dependence (3.23).

tion $q(H)$. The inclusion of such a bounded functional response of the predator into (3.22), the density-independent model of predation, yields

$$\frac{dH}{dt} = rH - q(H)P$$
$$\frac{dP}{dt} = -dP + fq(H)P. \tag{3.24}$$

A simple example for a response function is the so-called disk equation, which is also known as Holling's Type II functional response (Figure 3.5; cf. Begon et al. (1986)):

$$q(H) = \frac{cH}{k+H}. \tag{3.25}$$

k is a positive constant. This modification destabilizes the system, leading either to an uncontrolled growth of the prey population and the extinction of the predator (if $d/f > c$; no positive equilibrium), or to divergent oscillations around the fixed point, as illustrated on Figure 3.6. An intuitive explanation of this result is that – due to its physiological limitations expressed in the functional response function – the predator population is less efficient in controlling its prey at high prey densities, and it is more efficient when the prey is rare.

The inclusion of a bounded functional response transforms the unconditional global stability of the density-dependent model (3.23) as well. What stability type the fixed point admits in this case depends on the parameters of the modified model

$$\frac{dH}{dt} = rH\left(1 - \frac{H}{K}\right) - q(H)P$$
$$\frac{dP}{dt} = -d(1+gP)P + fq(H)P. \tag{3.26}$$

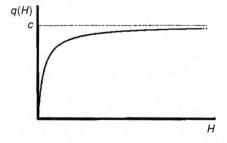

Figure 3.5 Per predator predation rate as a function of prey density: the Holling Type II functional response (3.25).

82 Population dynamics in patchy environments

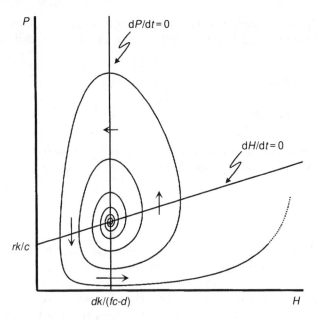

Figure 3.6 The destabilizing effect of Holling Type II functional response on the non-spatial Lotka–Volterra model of predation.

With only prey density dependence ($g = 0$) and a Holling Type II functional response (3.25) postulated, for example, the still unique positive equilibrium can be globally stable, neutrally stable, or locally unstable with a stable limit cycle around it, depending on the position of the predator zero isocline relative to the hump of the prey zero isocline (cf. Hofbauer and Sigmund, 1988) (Figure 3.7).

Predator–prey patch-abundance models have been extensively studied by many (e.g., Chewning, 1975; Comins and Blatt, 1977; Freedman and Waltman, 1977; Gurney and Nisbet, 1978; McMurtrie, 1978; So, 1979; Butler and Waltman, 1981; Crowley, 1981; Fujita, 1983; Pickett and White, 1985; Sabelis and Laane, 1986; Diekmann *et al.*, 1988; Hastings, 1991a; Sabelis *et al.*, 1991; Jansen and Sabelis, 1992; Vandermeer, 1993; Jansen, 1994, 1995) from many different aspects. The following sections present a selection of some interesting results from this ramified research area.

3.4.1 Diffusive coupling of identical predator–prey patches

The simplest possible step towards the spatiotemporal extension of the predator–prey Lotka–Volterra model (3.22) is to assume that the same system works on two identical patches, connected with linear dispersal:

Predation in patchy habitats

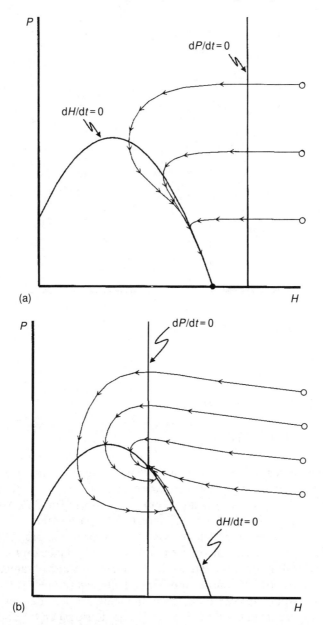

Figure 3.7 Stability properties of the non-spatial Lotka–Volterra model of predation with prey density dependence and a Holling Type II functional response of the predator. The outcome of the interaction depends on the efficiency of the predator in capturing and handling prey. (a) Stable fixed point with predator extinction; (b) stable coexistence; and (c) stable limit cycle.

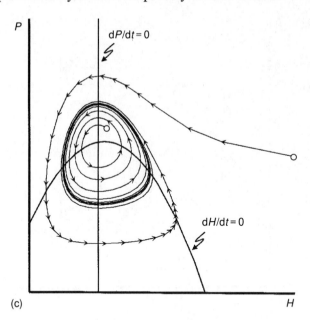

(c)

Figure 3.7 Continued

$$\frac{dH_i}{dt} = rH_i - H_iP_i + e_h(H_j - H_i)$$
$$\frac{dP_i}{dt} = -dP_i + H_iP_i + e_p(P_j - P_i)$$
, $i, j = 1, 2\ i \neq j$. (3.27)

All the parameters of this model are independent of locality, meaning that the patches offer exactly the same abiotic environment for both species on both patches. The prey-to-predator conversion factor (f in (3.22)) is assumed to be one for the sake of simplicity; the rest of the parameters are rescaled so that $c = 1$. e_h and e_p are the emigration rates for prey and predator, respectively; dispersal is conservative, without loss of abundance on the way.

Jansen (1994, 1995) studied this model numerically and obtained interesting, and also somewhat counter-intuitive results regarding its stability properties. One would expect that, since the patches are identical and net immigration occurs in the direction of the less dense patch for both species, what dispersal does is synchronize the two patches; in the limit, the synchronous system would behave like a single-patch model. On the other hand, this is what actually happens: the system always approaches a synchronous orbit around the fixed point. However, the synchronous periodic orbits of (3.27), which are obviously neutrally stable against spatially uniform perturbations, just like those of (3.22), might react rather differently to small heterogeneous perturbations.

Let us focus on the local stability of synchronous periodic orbits of (3.27).

The spatially uniform system is equivalent to the non-spatial model (3.22) (Figure 3.3), since the migration terms cancel with $H_1 = H_2$ and $P_1 = P_2$. The closed orbits of the synchronous system can be unambiguously labelled by their period lengths, τ (the duration of a complete cycle of the oscillation), because τ is a monotonically increasing function of amplitude (Waldvogel, 1986). Jansen (1994, 1995) showed that for any parameter combination (r, d, e_h, e_p) of 3.27, there exists a critical synchronous periodic orbit (of a period length τ_{max}) delimiting a 'stable disque' on the phase plane. All synchronous orbits **outside** the stable disque (that is, orbits with $\tau > \tau_{max}$) are locally unstable in the sense that after an arbitrarily small **heterogeneous** perturbation, the system reduces its period length (and thus also its amplitude) and gradually approaches another synchronous periodic solution **inside** the disque (Figure 3.8). Synchronous solutions **within** the disque are stable in the sense that after a small heterogeneous perturbation, the system approaches a synchronous orbit which is still inside the stable disque. In other words, the stable disque attracts all heterogeneously perturbed synchronous solutions. Figure 3.9 shows the local stability regions of the parameter space of (3.27), as a function of the dispersal rates e_h and e_p, with the parameters r and d kept constant. In Figure 3.9, each pair of dispersal rates (e_h, e_p) sits on exactly one stability contour curve, which corresponds to the period length τ_{max} (thus, also the amplitude) of the boundary orbit of the stable

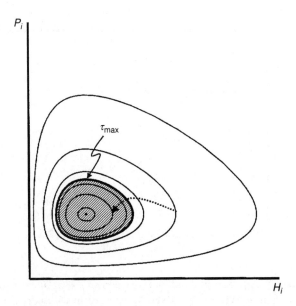

Figure 3.8 The stable disque of two identical predator–prey patches coupled by dispersal (system (3.27)) for the parameter set (r, d, e_h, e_p). Synchronous orbits outside the stable disque converge into the disque after a small heterogeneous perturbation; perturbed synchronous solutions within the disque do not diverge outside.

disque for those dispersal rates. The stable disque attracts all asynchronous trajectories ultimately; therefore the smaller the disque is, the more stable the system might be considered. Figure 3.9 shows clearly that for the stable disque to be relatively small, the prey should disperse weakly, whereas the predator's dispersal rate should be intermediate. Figure 3.10 is a numerical solution of (3.27), starting with a small perturbation of a synchronous orbit outside the stable disque. V.A.A. Jansen (unpublished results) has proven that essentially the same result can be transferred to cases with any number of identical Lotka–Volterra predator–prey patches; the more habitat patches we consider, the stronger the stabilizing effect turns out to be.

The moral of this study is straightforward: it tells us that the dispersive coupling of even environmentally identical predator–prey systems can be stabilizing in the sense that high-amplitude synchronous oscillations are damped to smaller amplitude ones; the effect is most pronounced if the prey population is a weak disperser and the predator is an intermediate disperser.

3.4.2 Dispersal asymmetry and stability in a two-patch Lotka–Volterra model

The synchronous solutions of (3.27) are still neutrally stable against uniform perturbations, the real part of the two conjugate dominant eigenvalues of its

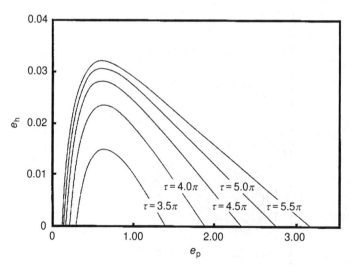

Figure 3.9 The effect of prey and predator dispersiveness (e_h and e_p) on the stability of synchronous solutions to (3.27). A synchronous orbit with period τ or greater is unstable for (e_h, e_p) parameter pairs **below** the graph corresponding to τ (it converges to another synchronous solution with a period smaller than τ, that is, into the corresponding stable disque) (after Jansen (1994)).

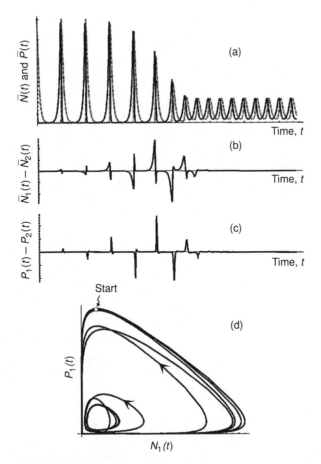

Figure 3.10 The way into the stable disque. A numerical solution of (3.27) with parameters $d = r = 1$, $e_h = 0$ and $e_p = 0.7$, starting from a slightly perturbed high-amplitude synchronous orbit. (a) The average abundances of prey (solid line) and predator (dotted line) over time on the two patches: $\bar{N}(t) = [N_1(t) + N_2(t)]/2$, and $\bar{P}(t) = [P_1(t) + P_2(t)]/2$; (b) the difference in the abundances of prey on the two patches, over time; (c) the difference in the abundances of predator on the two patches, over time; and (d) phase portrait of the dynamics of prey and predator on patch 1 (after Jansen (1994)).

Jacobian matrix being zero. A simple modification of the model leading to the local stability of the interior fixed point, that is, making the real part of the dominant eigenvalue negative, is the assumption of environmental differences between the two patches, affecting either within-patch dynamics, or dispersal. Consider the simple two-patch model studied by Murdoch *et al.* (1992) and Godfray and Pacala (1992):

$$\frac{dH_i}{dt} = r_i H_i - c_i H_i P_i - e_h H_i + z_i e_h (H_1 + H_2)$$
$$\frac{dP_i}{dt} = f c_i H_i P_i - d_i P_i - e_p P_i + q_i e_p (P_1 + P_2)$$
, $i = (1, 2)$ (3.28)

The notation in (3.28) is consistent with that of (3.22) and (3.27); e_h and e_p are the patch-independent constant emigration rates for prey and predator, respectively. The immigration terms (the last one in both equations) define the way emigrants are redistributed between the patches. z_i and q_i are fractions of immigrants from the pool of dispersers for prey and predator, with $0 \leq z_1 + z_2 \leq 1$ and $0 \leq q_1 + q_2 \leq 1$. If the fractions add up to one for any one of the species, dispersion is conservative regarding that species, otherwise a part of the dispersers gets lost.

Equation (3.28) is similar to (3.27), except that it allows for spatial environmental differences, and also for asymmetry in dispersion, characterized by $z_1 \neq z_2$ and/or $q_1 \neq q_2$. Dispersal asymmetry means unequal, and – for constant z and q values – also density-independent redistribution of dispersers among the patches. Equation (3.28) admits a unique positive fixed point, like (3.27). If all the parameters are the same for both species on both patches, including $z_1 = z_2 = 0.5$ and $q_1 = q_2 = 0.5$, that is, in case of symmetrical dispersal, the patches also become synchronized in the long run, even if they start from different abundance states. Once synchronized, the two patches behave like a single one, that is, (3.28) becomes (3.22), and the system is neutrally stable.

Murdoch and Oaten (1975) and Murdoch et al. (1992) show that spatial asymmetry in the parameters leads to stabilization of the fixed point usually, but instability can also be obtained in some cases. In particular, for a system without predator movement between patches ($e_p = 0$), dispersion asymmetry ($z_1 \neq z_2$) of the prey alone will stabilize the interaction, as the numerical simulation example on Figure 3.11 demonstrates. The direct cause of the stabilizing effect is the asynchrony of the two patches regarding both prey and predator cycles, which in turn arises from asymmetric dispersion.

3.4.3 Aggregation and stability in a two-patch environment

The effect of aggregation on stability is a permanent bother of optimal foraging theory and the population dynamics of host–parasitoid and predator–prey interactions (e.g., Allen, 1975; Cook and Hubbard, 1977; Munster-Svenson and Nachman, 1978; Comins and Hassell, 1979; Crowley, 1981; Lessels, 1985; Taylor, 1988; Bernstein et al, 1988, 1991; Reeve, 1988, 1990; Ives, 1992; Mangel and Roitberg, 1992; Godfray, 1994; Godfray et al., 1994). The problem is how the stability properties of a predator–prey (or host–parasitoid) model are affected, if the prey or the predator (or both) tend to congregate in patches that offer the best biotic milieu for them.

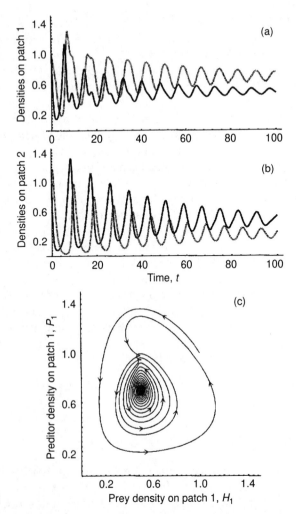

Figure 3.11 The stabilizing effect of the aggregation of dispersing prey in a two-patch predator–prey Lotka–Volterra system (3.28). Parameter values: $f = 1$, $e_p = 0$, $e_h = 0.5$, and $r_i = 1$, $c_i = 2$, $d_i = 1$ for $i = 1, 2$. The only difference between the patches is in prey aggregation: $z_1 = 0.9$, $z_2 = 0.1$. All initial densities were 1.0. (a) Prey (solid line) and predator (dotted line) densities on patch 1; (b) prey (solid line) and predator (dotted line) densities on patch 2; and (c) phase portrait of the local dynamics on patch 1.

From the prey's point of view, this means avoiding both conspecifics and predators as much as possible, whereas for the predator it is best to avoid patches with a high abundance of conspecifics, but it is advantageous to seek patches with high prey density. The answers to the question are rather sensitively dependent on the details of the actual model applied; therefore the conclusions suggested by different authors are often contradictory. The aggregation versus stability issue generated a fairly vigorous and witty controversy in *American Naturalist* between Godfray and Pacala (1992) and Murdoch et al. (1992), stemming from findings from an earlier model by Murdoch and Stewart-Oaten (1989). This model led the authors to a conclusion somewhat against the conventional wisdom of optimal foraging theory, stating that if a predator behaves optimally in the above sense, it will normally destabilize a temporally continuous system within the realistic part of its parameter space. Godfray and Pacala (1992) challenged this conclusion on the basis of a short simulation analysis of a system equivalent to (3.28), insisting that the results of the Murdoch and Stewart-Oaten (1989) model obtain from (3.28) is an unrealistic limit case with infinitely fast movement by both prey and predator. Murdoch et al. replied with an extensive simulation analysis of the same system, which is worth studying here in some detail, both because of its methodology and its conclusions.

Aggregation in favourable biotic environments means density-dependent habitat choice in general, which can be implemented into (3.28) as a specific form of density dependence to the immigration parameters z_i and q_i. Maintaining the assumption of conservative dispersion and ignoring predator effect on the dispersion of both predator and prey, a reasonable density-dependent option for z_i, the fraction of dispersive prey arriving at patch i, might be

$$z_i = \frac{(H_j)^u}{(H_1)^u + (H_2)^u}, \quad i, j = (1, 2), \; i \neq j \qquad (3.29)$$

where the exponent $u \geq 0$ determines the strength of density dependence. $u = 0$ represents complete density independence in prey migration, whereas increasing u results in increasingly preferential migration towards the patch with fewer prey. q_i, the fraction of dispersive predator landing on patch i, can be defined analogously as

$$q_i = \frac{(H_i)^u}{(H_1)^u + (H_2)^u}, \quad i = (1, 2) \qquad (3.30)$$

with preferential movement towards the patch maintaining a larger prey population as u exceeds 0.

Murdoch et al. (1992) first establish differences in the stability properties of (3.28) as a result of aggregation separately for prey and predator, in order to avoid confusion due to the superposition of unknown elementary effects; then they inquire into the joint effect of prey and predator aggregation for some specific situations.

The null model for the study of prey aggregation impact is (3.28), with no predator movement ($e_p = 0$) and with spatial asymmetry only in prey birth rate ($r_1 \neq r_2$) assumed. This system admits a unique interior fixed point, which is locally asymptotically stable, the real part of the leading eigenvalue of its Jacobian matrix being always negative (cf. Appendix B) (see Figure 3.12). Now substituting (3.29) for z_i, we ask how the local stability of the fixed point changes with increasing density dependence (increasing u) in prey habitat choice, and what mechanisms might be directly responsible for the change.

The problem requires a semianalytical treatment, including calculations to determine the real part of the leading eigenvalue as a measure of stability, covariance of same-instant prey abundances on the two patches to measure prey asynchrony, and the density dependence of realized prey immigration. Figure 3.13a shows that if a large part of the prey population is always ready to disperse (that is, e_h is sufficiently large), density-dependent habitat choice by the prey is strictly destabilizing in terms of decreasing the absolute magnitude of the real part of the leading eigenvalue, but not in terms of changing its sign to positive. The system remains in fact locally stable, but its resilience (the speed of return to the fixed point when perturbed, cf. Box

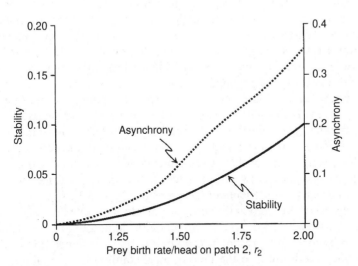

Figure 3.12 The dependence of the stability and the local asynchrony of the two-patch predator–prey Lotka–Volterra system (3.28) on the difference between prey growth rates (r_i) on the two patches. Parameters as in Figure 3.11, except for r_2 (ranging from 1.0 to 2.0), $e_h = 1$ and $z_1 = z_2 = 0.5$. 'Stability' is measured by minus the real part of the leading eigenvalue of the Jacobian matrix of (3.28); local asynchrony is $1 - \rho$, where ρ is the correlation between prey densities on the two patches, calculated from time 15 onward. Both stability and asynchrony increases with the local difference in prey growth rates (after Murdoch *et al.* (1992)).

3.3) decreases with density-dependent prey movement introduced. The background mechanism again is to do with the asynchrony of prey on the two patches, which is obviously ruined if there are many dispersers seeking scarcely populated habitats. Note that this finding is in good agreement with the conclusion of Jansen's model (Section 3.4.1) suggesting that intensive migration of the prey population towards the less crowded habitat increases the size of the stable disque (cf. Figure 3.9), that is, it decreases stability.

The situation is somewhat different if the prey emigration rate, e_h, is smaller. Less intensive prey dispersion makes slightly density-dependent habitat choice advantageous, but strong density dependence (large u) is again adverse in terms of resilience (Figure 3.13b). The mechanism behind this result is surely much more complicated than asynchrony alone, however, and it is by no means transparent. Asynchrony is obviously part of the story, but the immediate cause of the hump on the stability curve is probably also connected to the local efficiency of prey in controlling predator and vice versa.

If we turn the dispersal relations to their opposites, so that it is the predator population alone which is allowed to disperse preferentially (according to (3.30)) between the patches, while the prey organisms are strictly bound to the patch they were born in ($e_h = 0$), it is again only the resilience of the null model that is affected by aggregation; the effect is also dependent on the strength of aggregation. Slight predator preference for dense prey patches tends to increase the resilience of the interaction, but stronger preference reverts the tendency, driving the locally stable null system towards neutral stability (Figure 3.14a). The underlying mechanism might be two-fold, one effect superimposed on the other: asynchrony arising from predator movement independent of conspecific abundance increases resilience, whereas the synchronizing effect of more intensive predation on the more abundant prey population destabilizes the interaction. Which of these effects dominates the process depends on the strength of predator aggregation. This result does not change qualitatively if we also assume that the prey population disperses between the patches, provided prey dispersion is symmetric and density-independent ($z_1 = z_2 = 0.5$; Figure 3.14b).

If dispersive prey tends to aggregate on the poorer patch regardless of local density ($r_1 > r_2$, $z_1 < z_2$) and the predator population aggregates on the patch with more prey present, it is no longer only resilience that is affected as u is increased, but also the type of local stability that the interior fixed point admits. Namely, sufficient predator preference towards the more abundant prey patch destabilizes the interaction in terms of changing the sign of the leading eigenvalue to positive (Figure 3.14c). This is certainly a result of the self-synchronizing mechanism of prey due to disperser aggregation on the patch with smaller prey growth rate.

As a conclusion, one can say that preferential habitat choice in itself is neither stabilizing nor destabilizing in general; the effect is strongly context-

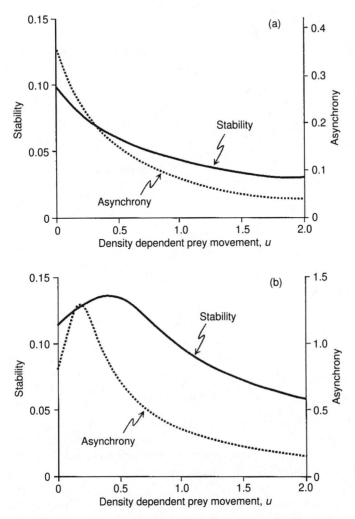

Figure 3.13 Stability and asynchrony of the (3.28)–(3.30) system as a function of the density dependence measure u of prey movement. Other parameters (a) as in Figure 3.12, but $r_1 = 2.0$; (b) as in (a), but $e_h = 0.5$ (after Murdoch *et al.* (1992)).

dependent. Essentially this is the conclusion also arrived at later by Rohani *et al.* (1994), based on a detailed stage-structured patch model of host–parasitoid interaction. Prey aggregation on less crowded habitats might stabilize, but it can also destabilize the interaction, and predator aggregation effect seems to be dependent on the strength of predator preference towards the more abundant prey patch. A fairly general immediate determinant of regional stability in predator–prey patch dynamics seems to be the

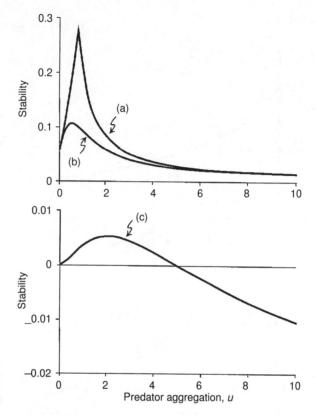

Figure 3.14 The effects of predator aggregation on the stability of the (3.28)–(3.30) system. (a) No prey movement ($e_h = 0$), maximum predator movement ($e_p = 1$), other parameters as in Figure 3.11; (b) symmetric prey movement added: $e_h = 0.5$ and $z_1 = z_2 = 0.5$, other parameters as in (a); and (c) prey seek the 'poorer' patch: $r_1 = 1$, $r_2 = 0.7$, $z_1 = 0.3$, $z_2 = 0.7$, $c_1 = 1.0$, $c_2 = 0.7$, $f = 1$, $e_h = 1.7$, $e_p = 0.85$. Strong enough predator aggregation leads to instability in this case. Note the change of scale in (c)! (after Murdoch *et al.* (1992)).

asynchrony of local dynamics. In fact, the dynamical relevance of most assumptions can be judged qualitatively through their impact on asynchrony, if the latter is transparent enough.

3.4.4 The effects of predator mobility and delayed functional response

A number of theoretical questions pertaining to simple and rather specific dispersive predator–prey systems have been answered by using analytically tractable, but also technically more involved models. A brief synopsis of such results is given below without details.

Freedman and Takeuchi (1989a) study an n-patch model with linear prey dispersal but unlimited predator movement (that is, the habitat is patchy for the prey, but not for the predator). Prey density dependence and predator functional response are defined as arbitrary functions that satisfy reasonable and not very rigorous criteria. The authors derive conditions for the persistence of the predator in relation to its death rate and functional response. For the special case of two patches, Kuang and Takeuchi (1994) were able to establish precise conditions for the existence and the stability of positive equilibria, and conclude that both increasing or decreasing the intensity of prey dispersal might destabilize initially stable systems of this kind. A more realistic version of this two-patch model with both prey and predator dispersing according to a linear migration mechanism is analysed by Freedman and Takeuchi (1989b), who show that if the predator is persistent, the system admits an asymptotically stable fixed point inside the positive orthant of the state space.

Beretta *et al.* (1987) postulated the delayed functional response of the predator to prey density in a specifically constrained two-patch model, assuming that the prey-limited predator is restricted to one of the patches where prey is predator-limited, and on the other, predator-free patch, prey is self-limited (density dependent). Prey can disperse between the 'predation-free, low food' and the 'predation risk, abundant food' patches in a diffusive way. Beretta *et al.* (1987) found that the delayed response to prey density destabilizes this system. This result might not seem surprising intuitively, time delay being in most non-spatial systems destabilizing. An interesting – and still open – question, however, is what would be the effect of delayed **migration**, instead of the delay in functional response? Knowing that local asynchrony has a general stabilizing effect on the regional dynamics of patch models, the realistic assumption of a time delay in dispersion (that is, in cases when emigrants leaving a patch need time to land on another) might be even advantageous for regional stability, possibly inducing local asynchrony between the patches. This question is worthy of some formal analysis in the future.

3.5 CHAOTIC DYNAMICS OF SINGLE-SPECIES SYSTEMS IN PATCHY ENVIRONMENTS

The assumption of time delay effects makes even single-species non-spatial systems difficult to analyse by analytical means, but – besides possible improvements in biological feasibility – it allows a wider range of the types of dynamical behaviour to occur across the parameter space as compared to what the corresponding non-delayed single-species models can produce. Discrete time models of population dynamics are to some extent analogous to continuous time delayed systems in the sense that also in discrete time models it is a state in the past (in time unit $t - 1$) that determines what happens now (at time t). Time lags in regulatory feedback processes such as density-dependent births and deaths tend to destroy stability, for the quite obvious reason that the lagged system responds to earlier stimuli instead of the actual ones, which might have changed radically in the meantime. This is the reason why time-discrete, difference-equation versions of the logistic model, for example, show an astonishingly colourful range of stability types from monotonic damping onto an asymptotically stable positive fixed point through oscillatory damping and stable limit cycle to chaos (cf. May, 1976), in contrast to the unconditional asymptotic stability of their continuous equivalent, the Pearl–Verhulst model.

3.5.1 Migration against chaos: dispersion and stability in coupled maps

Even single-species systems can produce very complex dynamics in a patchy environment setting, as demonstrated simultaneously and independently by Hastings (1993) and Gyllenberg et al. (1993) using the same quadratic map version of the discrete logistic model on two dispersively coupled patches. The local dynamics of the two populations are

$$\tilde{x}_t = r_x x_t (1 - x_t)$$
$$\tilde{y}_t = r_y y_t (1 - y_t), \qquad (3.31)$$

where x_t and y_t are the population abundances on the two patches before reproduction, $\tilde{x}_t$ and $\tilde{y}_t$ are those after reproduction but before dispersion, r_x and r_y are the corresponding local reproduction rates without the linear density effect, itself represented by the bracketed terms. Note that the 'carrying capacity' is normalized to one for both patches. Dispersion is simply linear: a fraction D of the reproduced population emigrates without loss to the other patch immediately:

$$x_{t+1} = \tilde{x}_t + D(\tilde{y}_t - \tilde{x}_t)$$
$$y_{t+1} = \tilde{y}_t + D(\tilde{x}_t - \tilde{y}_t). \qquad (3.32)$$

The assumption that dispersers are redistributed randomly between the patches constrains D under 0.5, since even if the whole population emigrates, half of it returns to the patch of its origin. Substituting (3.31) into (3.32) yields

$$x_{t+1} = r_x (1 - D) x_t (1 - x_t) + r_y D y_t (1 - y_t)$$
$$y_{t+1} = r_y (1 - D) y_t (1 - y_t) + r_x D x_t (1 - x_t), \qquad (3.33)$$

the model of first reproducing, then dispersing populations of a single species on two patches.

Chaotic dynamics of single-species systems

The quadratic map representation (3.31) of local dynamics has the strange property that the state variable can 'overshoot' the zero abundance line, thus taking negative population abundance values if the reproduction rate is larger than four, or if $x_t > 1$. Therefore the justification for its use might be mathematical convenience in the first place, but Hastings (1993) also notes that simulating a modified version of (3.33) with a more realistic (bounded from below at zero) substitute of (3.31),

$$\tilde{x}_t = x_t \, e^{r(1-x_t)} \tag{3.34}$$

(cf. May, 1977) does not make a qualitative difference in the behaviour of the trajectories. Another feasible form would be

$$\tilde{x}_t = r \frac{x_t}{(1 + ax_t)^\beta}, \tag{3.35}$$

where a and β are constants defining the strength of density dependence in local growth, but the fact that it has three parameters as opposed to the one in (3.31) makes it inconvenient to use if analytical tractability is an important point.

Anyway, (3.31) is still an acceptable option at least as a toy model if values of r are kept below four and no initial conditions are allowed above the carrying capacity, since in that case all the orbits of (3.33) remain within the invariant set $I^2 = [0, 1] \times [0, 1]$ of the state space.

Equation (3.33) shows extremely complicated dynamics within its parameter space even if the patches are identical, so that $r_x = r_y = r$. The most interesting part of its $r - D$ plane is the interval $3 \leq r < 4$ and $0 \leq D \leq 0.25$. Any parameter combination with a low rate of reproduction (namely, $r < 3$) leads to a uniform, asymptotically stable fixed point behaviour, which is the only outcome of the continuous time equivalent of the model, regardless of parameter values. On the other hand, if the patches are strongly coupled through very intensive dispersion (specifically, if $D > 0.25$), they become synchronized so that the two populations unite to form a single one in effect, and the spatial aspect no longer plays a role in the dynamics.

Within the relevant part of the parameter space, the repertoire of possible dynamics turns out to be really colourful; it seems hopeless to explore it completely either analytically or numerically. The single-patch (non-spatial) version has been extensively studied as a prototype chaotic system, and although even that is very complicated, its behaviour is fairly completely known (cf. Devaney, 1986). To illustrate the possible complications and the beauty of this seemingly innocent system, let us see very briefly how that non-spatial system behaves as r, its only parameter, is increased above three. In passing $r = 3$, the fixed point loses stability (but not existence), and a stable limit cycle of period two emerges, which loses stability at about $r \cong 3.45$, whereby a stable period-four cycle takes over; that in turn is succeeded by a period-eight cycle, and so on. This infinite series of period-doubling bifurcations continues with an increasing density along r up to a finite value (approximately $r = 3.57$), where an aperiodic, fractal attractor arises as the limit of this so-called Feigenbaum, or 'direct cascade' scenario. With r increasing further, orbits of periods of powers other than two emerge according to Sharkovski's order (cf. Devaney, 1986), which implies the existence of chaos as well. Whether the chaotic behaviour is only transient or if it is asymptotic for almost all initial conditions depends on r in an intricate way. Asymptotically periodic behaviour, together with transient chaos, occurs within intervals of r. These 'periodic windows' are separated by values of r at which the system admits a strange (chaotic) attractor, which may or may not be fractal. Such values of r constitute a fractal set of positive measure.

What new information does the dispersive coupling of two such local populations add to this picture? Only a very tentative answer can be given to this question, because much less is known about the system (3.33) than its non-spatial 'pre-image'. Nevertheless, some important conclusions emerge from analytical and simulation studies done by Hastings (1993) and Gyllenberg et al. (1993).

First, the stable periodic solutions (limit cycles) of the coupled system (3.33) can be either in-phase, meaning that the abundances oscillate synchronously on the two patches, or out-of-phase, with the local abundances following different periodic dynamics (Figure 3.15).

Second, the in-phase and the different out-of-phase solutions may or may not be stable **simultaneously**, meaning that there is a possibility for multiple attractors to exist in this case, unlike in the non-spatial model which admits a single attractor for a given value of r. This possibility is illustrated in Figure 3.16, which is a joint stability map of the in-phase and the out-of-phase period-two orbits within the relevant part of the parameter space. Note that a similar map could be created and overlayed on this one (at least in principle) for orbits of higher periods as well, so that there might be parts of the parameter space admitting even more attractors. There are parts within the stability region of the out-of-phase period-two solution, for example, which also correspond to stable out-of-phase period-four orbits. The overlap of stability regions, that is, the coexistence of different periodic attractors within certain domains of the parameter space (Figure 3.16), implies that the phase plane I^2 consists of separate basins of attraction. Which stable orbit absorbs a certain trajectory depends on the initial local abundances. The geometry of the basins of attraction is illustrated in Figure 3.17a for the case with an in-phase and an out-of-phase period-two orbit. Fractal basins of attraction emerge within parts of the parameter space admitting at least two different stable out-of-phase orbits (Figure 3.17b). The fractal nature of the basins of attraction means that the ultimate spatiotemporal abundance pattern of the two-patch system is very sensitively dependent on the initial conditions – in fact, it is almost impossible to predict which periodic attractor the system approaches in this case.

The third interesting, and from our viewpoint also very important, feature of the system (3.33) is that dispersion tends to stabilize the out-of-phase

Chaotic dynamics of single-species systems

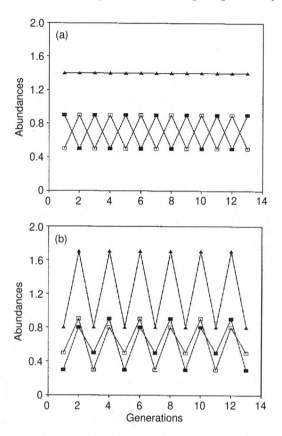

Figure 3.15 Out of phase (that is, locally asynchronous) solutions of the coupled quadratic map system (3.33) with parameters $r = 3.8$ and $D = 0.15$. (a) The period-two solution with constant total abundance; (b) the period-four solution, where the total population oscillates with period two. Open and filled squares: the two sub-population abundances; filled triangles: total abundances (after Hastings (1993)).

period-two orbits against chaos. It does this rather effectively, since there is a considerable, and also rather realistic, range of dispersal which maintains the stability of the two-periodic orbits even within the chaotic parameter domain (close to $r = 4$) of the non-spatial model (regions 2 and 4 on Figure 3.16). Gonzalez-Andujar and Perry (1993) obtained essentially the same result (without the details on periodicity) by a simple numerical study of the same model.

3.5.2 Coupled map lattices: the multipatch extension of the coupled logistic model

The conclusion that the dispersive coupling of unpredictably oscillating local populations has a stabilizing effect on the multipatch system as a

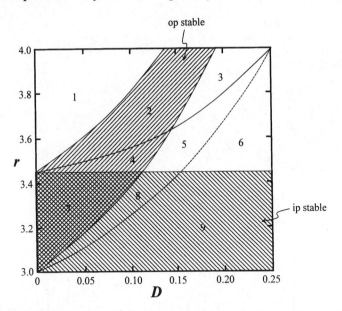

Figure 3.16 Stability map of the synchronous (in phase; ip) and the asynchronous (out of phase; op) period-two orbits of (3.33). Numbered regions of the map: 1, ip source, op source; 2, ip source, op sink; 3, ip source, op saddle; 4, ip saddle, op sink; 5, ip saddle, op saddle; 6, ip saddle, op does not exist; 7, ip sink, op sink; 8, ip sink, op saddle; 9, ip sink, op does not exist. Right-hatched region: op is stable; left-hatched region: ip is stable; cross-hatched region: both op and ip are stable (after Gyllenberg *et al.* (1993)).

whole, might be the basis of an explanation for the fact that chaotic or chaotic-like dynamics appear to be quite rare in nature, even among arthropods that are known to have a large fecundity (that is, large r) and a seasonal (temporally discrete) dynamics, indicating at least a propensity for chaotic behaviour (cf. Turchin, 1991; Hassell *et al.*, 1991; Godfray and Grenfell, 1993; Vandermeer, 1993). This result is rather intuitive, and one expects it to apply to the corresponding multipatch model as well. The situation becomes somewhat more complicated with an increased number of patches, however.

Multipatch coupled discrete logistic systems are called 'coupled map lattices' (CMLs) in the recent literature of both physics and biology; it is a topic of common interest with diverse fields of potential applications, ranging from spin-glass modelling through neural networks to population biology. In a population dynamical setting, a CML is the implementation of a system like (3.33) within a grid of patches, so that the dispersal connectivity between pairs of patches is established either according to spatial neighbourhood relations, or in an arbitrary manner. The grid can be of any number of spatial dimensions, the most reasonable options obviously being

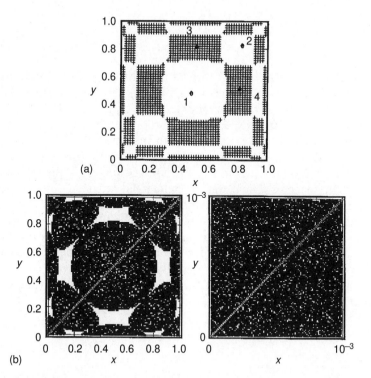

Figure 3.17 (a) Basins of attraction for the stable period-two orbits of (3.33), with parameters $r = 3.3$ and $D = 0.035$ (these fall in region 7 of Figure 3.17). The white area is the basin of attraction of the in-phase (ip) solution (represented by points 1 and 2; the cross-hatched region is that of the stable out-of-phase (op) orbit (points 3 and 4) (after Gyllenberg *et al.* (1993)). (b) Fractal basins of attraction of the period-two (white) and the period-four (black) out-of-phase solutions of (3.33) with parameters $r = 3.8$ and $D = 0.15$. Initial conditions with X-marked white blocks are attracted by other – typically chaotic – solutions. Left: the whole range of initial conditions; right: a small fraction of the phase plane with very small initial abundances (after Hastings (1993)).

one to three in population dynamics – although the use of higher-dimensional (hypercubic) lattices can also be argued for (cf. Chaté and Manneville, 1988, 1992), if the dimensionality of the lattice only measures patch connectivity or it defines the connectivity structure in a concise way, and is not meant to represent any real spatial arrangement. Neither the topology of the connectivity map nor spatial dimensionality affects the tractability of a CML significantly: it resists analytical treatment greatly anyway. The only practicable method is numerical investigation by computer.

A simple, but fairly general, two-dimensional CML with linear dispersal terms can be

$$x'_{ij} = (1-D_{ij})f(x_{ij}) + \sum_{kl \in \Theta(ij)} f(x_{kl})\frac{D_{kl}}{N_{\Theta(kl)}}, \qquad (3.36)$$

where x_{ij} is the abundance of the local population on patch ij (i and j are grid site indices) at time t, x'_{ij} is that at time $t+1$; $f(x_{ij})$ is the map function defining within-patch dynamics, which might be of the type (3.31), (3.34) or (3.35), for example; D_{ij} is the dispersed proportion of the local population ij. Dispersers are distributed evenly among the patches belonging to the recipient patch set $\Theta(ij)$ (which includes all the patches getting dispersers from the focal patch ij). $N_{\Theta(ij)}$ is the number of patches belonging to the recipient set of ij. By far the most usual way to define the recipient set is to simply consider a certain neighbourhood of the focal patch, but this is not a necessity at all.

Chaté and Manneville (1988) use a d-dimensional, but otherwise simplified version of (3.36), with $D_i = 1$ for any patch i within the grid (complete dispersion), identical growth rates r on all patches, and the recipient set $\Theta(i)$ including the orthogonal nearest neighbours for each patch (patches are labelled by a single index below, because the spatial dimensionality of the system is not specified):

$$x'_i = \sum_{j \in \Theta(i)} rx_j(1-x_j)\frac{1}{N_{\Theta(j)}}. \qquad (3.37)$$

Note that $f(x_i)$ is the same map function as (3.31). The dispersion mechanism is such that each patch takes the abundance average of its $N_{\Theta(j)} = 2d$ orthogonal nearest neighbours after reproduction (or, equivalently, each patch distributes its whole local population evenly among the neighbouring patches).

Numerical simulations with this model lead to the following conclusions in more than one dimension (Chaté and Manneville, 1988, 1992): within the Feigenbaum region of the control parameter r, that is, for $3 < r \leq 3.57$, the lattice as a whole approaches the local dynamics of the system, that is, it converges to closed orbits with powers-of-two periods. Above the direct cascade region, within $3.57 \leq r < 4$, the lattice does not follow the single-patch dynamics any more, apart from the case of a uniform start; all patches produce similar chaotic-like time series in this interval, without signs of periodic windows being present, and the time series never synchronize. However, this does not necessarily lead to a noisy fixed point behaviour of the lattice average, which would nevertheless be a reasonable expectation: much more frequently the lattice average is a noisy oscillator instead (Figure 3.18), which proves to be unexpectedly robust against the addition of not very high-amplitude white noise, and also against changes in initial and boundary conditions.

Changes in the topology and the temporal structure of the connectivity map have been shown to be very important regarding the synchronization of

Chaotic dynamics of single-species systems

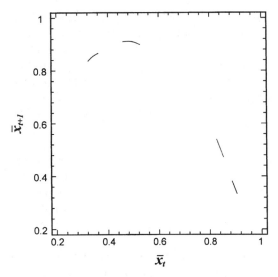

Figure 3.18 First return map of a five-dimensional coupled map lattice of size 35^5, with $r = 3.83$ (3.35). The lattice average of the local abundances admits a noisy period-four cycle.

local dynamics within the lattice (Chaté and Manneville, 1992). The first method to study these effects was to choose the recipient patch set $\Theta(i)$ from the lattice at random for each patch i, so that its size N_Θ was an arbitrary constant, independent of i. N_Θ represents the connectivity of the lattice. The arrangement of the edges of the connectivity graph might change through time, or else the random arrangement can be 'frozen'. If it is frozen, then the choice of edges is once forever, meaning that the connectivity graph is time invariant; otherwise it is rearranged using the same stochastic rule every time step.

The most important result of the simulations based on these assumptions is that synchronization of the lattice occurs both with increasing the connectivity N_Θ and with changing from the frozen random to the temporally changing random scenario of patch connectivity. The synchronizing effect is independent of r if N_Θ is large enough (for the frozen random case, the threshold is $N_\Theta = 7$; for the temporally changing case, it is $N_\Theta = 4$, above which the lattice gets synchronized regardless of the actual value of r). For smaller values of N_Θ, there exists a critical r below which synchronization does not occur.

The second, 'step-by-step delocalization' method was to change the edges of the connectivity graph of a nearest neighbour model so that a given number $H \leq N_{\Theta(i)}$ of the nearest neighbours for each patch i are substituted by a randomly chosen patch from the lattice. The choice may or may not be 'frozen' for the whole time series also in this case. Clearly both the delocalization (spatial extension) and the temporal variability of dispersal connections represent an increased measure of spatial mixing, that is, a loss in the spatial focusing and temporal memory of patch interactions.

Note that when the recipient set is the orthogonal nearest neighbourhood, the connectivity of the lattice increases with its dimensionality, given $N_{\Theta(i)} = 2d$ for an interior cell i. Increasing d has also an apparent spatial mixing effect if the connectivity map is projected onto a lower-dimensional lattice. This is a misleading impression, however, in the strict nearest neighbour case ($H = 0$), since the topological structure of the neighbourhood relations is preserved on such an 'unfolded' map; the projection onto a lower-dimensional lattice simply makes the neighbourhood relations more difficult to realize. When the neighbouring patches are gradually changed to randomly chosen ones, however, we might expect the spatial mixing effect to become more expressed, amplifying the connectivity effect as well.

In accordance with the expectations, the results so far reported on this scenario (Chaté and Manneville, 1992) show that increasing H, the non-local proportion of patch interactions, tends to synchronize the dynamics of the patches. First it alters the collective dynamics in the form of an added 'noise', masking the details of the nearest neighbour model outcomes, and then, with H approaching N_Θ, the lattice seems to synchronize as if it were a single patch. This effect, of course, appears to be even more pronounced with temporally changing connectivity graphs, whereby the topology of connections is not only random, but also changes from generation to generation.

3.5.3 Self-organized criticality defeats chaos in a coupled map lattice

Self-organized criticality is a recent concept in the physics of many-body systems (Bak et al., 1987, 1988); a concept that has since found its way into many other disciplines, including other branches of physics, economics and also population biology. Before details on its use in a population dynamical context are given, it might be helpful to circumscribe its meaning with the help of a typical physical example – even at a price of possibly being accused of starting the story too far from the focal subject.

Consider a sand pile being fed from above through a point-like source of slowly running sand. As the quantity of sand in the pile increases, the pile soon takes a conical shape, which remains steady in spite of the increase in its volume; in particular, it is the angle of the surface to the base of the cone that stays constant, and it is maintained by the following mechanism. As sand accumulates on the top of the pile, and thus the slope angle becomes steeper locally, the component of gravity force parallel to the slope exceeds the adhesion force among some particles, which consequently start rolling down. Their kinetic energy pushes many other particles out from the adhesive tie; these will also start rolling down, pushing other particles, and so on. This cascade mechanism generates 'avalanches' of different sizes on the surface, which may or may not run down to the base of the cone. The state of the sand pile is critical in the sense that its particles are very close to the balance between the slopeward gravity force component and adhesion force, so that very small perturbations might trigger avalanches of unpredictable sizes; it is only the size distribution of the avalanches which can be calculated for a given situation. The critical state is also self-maintaining: the return of the slope angle to its critical value is only a question of time after perturbations.

The organizing principle of the sand pile is the threshold behaviour of its particles. Analogous threshold-driven phenomena are well known in population dynamics, the most obvious examples of which would be cases of density-dependent emigration according to a step function. This implies density dependence only above a certain critical level of abundance, at which density stress simultaneously triggers the dispersal behaviour of many individuals, similar to the mass migration of certain insect species (just think of the well-known phenomenon of locust invasions as an example). Scheuring et al. (1993), Csilling et al. (1994), Jánosi (1994) and Scheuring and Jánosi (1996) consider this case in the framework of a two-dimensional CML model with local dynamics of the (3.35) type:

$$x'_{i,j} = r \frac{x_{i,j}}{(1+ax_{i,j})^{\beta}}. \tag{3.38}$$

Equation (3.38) may admit a stable fixed point, stable limit cycles or chaos, depending on the parameters r and β. Dispersion starts only if local abundance exceeds a certain threshold value k on at least one of the patches. It is an orthogonal nearest neighbour interaction, with the following rules applied:

$$\text{If } x_{i,j} > k \text{ then } x_{i,j} \to x_s, \text{ and } x_{i\pm 1, j\pm 1} \to x_{i\pm 1, j\pm 1} + D\frac{x_{i,j} - x_s}{4} \tag{3.39}$$

where x_s is a uniform subcritical abundance (downward threshold) at which dispersion ceases, and $0 < D \leq 1$ is the constant of within-lattice dissipation (a fraction $1 - D$ of the emigrants gets lost during dispersion). Equation (3.39) implies that the excess abundance above x_s, the downward threshold of dispersion switch-off, is distributed evenly among the four nearest neighbours of the focal patch. x_s is smaller than the upward threshold k of dispersion onset: $x_s < k$. Emigrants leaving a patch might increase the abundance within the neighbourhood over the critical level k, and thus trigger a dispersion cascade if the lattice is close to the critical state. Such 'dispersal avalanches' dissipate the temporary excess of abundance through the absorptive boundaries of the lattice.

Two important points should be made regarding the constants of the model in advance. First, simulation of the systems (3.38) and (3.39) shows that the measure of hysteresis in switching dispersion on and off, due to the strict inequality in this relation, does not affect the results of the model qualitatively, if $k - x_s$ is not unrealistically large. Second, also the actual value of the constant D of within-lattice dissipation did not matter in qualitative terms, and it was therefore kept at a value of one to make the system as simple as possible.

The imposition of the threshold criterion turns out to be of decisive importance regarding the dynamics on both the patch and the lattice level. Starting from the chaotic domain of (3.38), local dynamics tend to stabilize with decreasing k, from chaotic through noisy periodic to a noisy fixed point behaviour (Figure 3.19). One would expect the stabilizing effect of criticality to manifest itself clearly also on the lattice level, but this turns out not to be the case, at least not so clearly as on the local level. The average abundance of the uncoupled lattice (with an infinite threshold value k) behaves seemingly better than that of a coupled system with a moderate upward threshold of dispersion, the former producing a noisy fixed point, and never approaching the deadly zero abundance state, as opposed to the violent limit cycle oscillations of the latter (Figure 3.20). The limit cycle on the lattice level is obviously due to the synchronization of cycles emerging on the local level as a result of the threshold behaviour. The patch-level limit cycle shows up as a sharp peak in the middle of the power spectrum of the patch on Figure 3.19. With k increased further, the lattice average approaches a fixed point and a very low level of noise is superimposed, which is clearly a valuable property in terms of population persistence.

Although Figure 3.20 might suggest that it is disadvantageous to decrease k, the critical abundance level triggering emigration when it is large, the picture changes dramatically if one considers the effect of random local perturbations as well. In an uncoupled lattice, each patch spends a considerable part of its time close to the zero abundance state (see Figure 3.19a), which will absorb it sooner or later if a small amplitude white noise term is added to (3.38). In the non-dispersive $k = \infty$ case the local zero state is a sink, since no immigrants come to recolonize the patch. The population certainly goes extinct on the whole lattice in finite time, even though the lattice average of the undisturbed system seems to stay comfortably far away from zero (see Figure 3.20a), that is, the system is not robust against small local random perturbations. For the seemingly more vulnerable large-amplitude limit cycle case, small local perturbations cannot drive the population to extinction on the lattice level even if local extinctions do occur from time to time, since frequent 'dispersal avalanches' will soon recolonize the extinct patches. It is only global perturbation affecting all patches simultaneously that might drive the system to instantaneous extinction in this case. And it

Chaotic dynamics of single-species systems 107

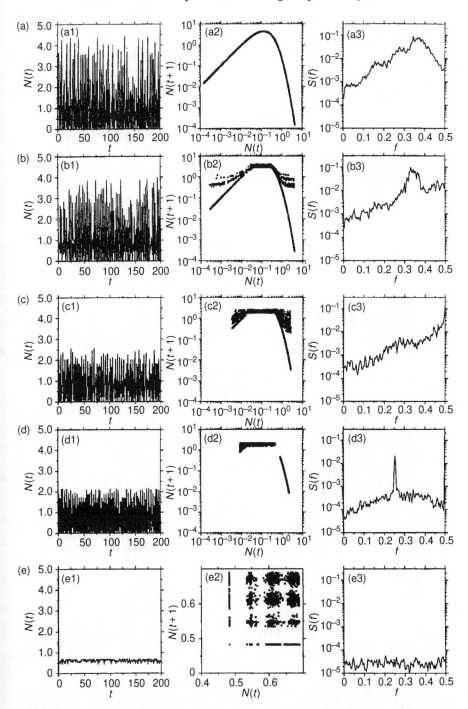

Figure 3.19 Time series, first return maps and power spectra of the local population density in a central site of a 16 by 16 lattice with different threshold parameter (k) values: (a) $k = \infty$ (no dispersion); (b) $k = 4.0$; (c) $k = 2.69$; (d) $k = 2.165$; and (e) $k = 0.69$. Note the change of scale in the first return map of (e)!

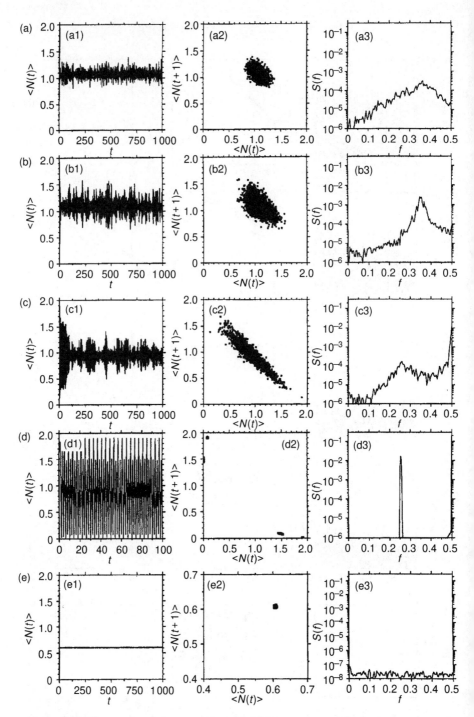

Figure 3.20 Time series, first return maps and power spectra of the lattice average of local population densities for the 16 by 16 lattice with different threshold parameter (k) values: (a) $k = \infty$ (no dispersion); (b) $k = 4.0$; (c) $k = 2.69$; (d) $k = 2.165$; (e) $k = 0.69$. Note the change of scale in the first return map of (e)!

is resistant even against global perturbations, if k is sufficiently small for the system to admit a slightly noisy fixed point.

In all, it seems to be true in general for this model that decreasing the critical abundance of triggering emigration has a beneficial effect on the propensity for global persistence in the locally chaotic parameter domain. Csilling et al. (1994) also show that the system approaches self-organized criticality (in terms of the shape of the dispersal avalanche-size distribution) as k decreases, so that the lattice-level persistence of the species is in fact tied to the self-organized criticality of the dynamics.

3.6 SUMMARY

The increased propensity of multipatch systems for persistence and coexistence, as compared with single-patch (non-spatial) systems of otherwise similar structure, is generally demonstrated by single-species and multispecies models alike. There are basically two sources of stability in patch-abundance models: environmental heterogeneity and local asynchrony. Environmental heterogeneity can act through different mechanisms, depending on which component of the dynamics – within-patch demography or interpatch dispersion – it affects. Demographic heterogeneity is usually stabilizing, because it opens the possibility for interacting species to get more or less separated in space according to their milieu preferences. If, on the other hand, biotic or abiotic milieu heterogeneities control the mechanism of dispersal, thus inducing the aggregation of dispersers on certain patches, the effect can be either stabilizing or destabilizing.

Even habitats consisting of environmentally identical patches can be safer for certain populations than a single-patch system governed by the same dynamics. Whatever type of biotic interaction is considered (competition, predation, parasitism, etc.) in an environmentally uniform patchwork habitat, the stabilizing effect may be usually attributed to the partial isolation of the subpopulations, which may asynchronize their local dynamics, thus making possible the recolonization of extinct habitat patches by immigration from abundant local populations. The effect of interpatch migration can be viewed as a specific type of mutualistic interaction among local populations, which helps the less abundant ones to increase at the expense of those with a safe, large density, far from any serious threat of local extinction. This is one aspect of the risk-spreading effect of patch-abundance dynamics. Also the resilience of patchily distributed populations can be stronger against heterogeneous perturbations, which represent a different mechanism of risk spreading.

The seasonal dynamics of density-dependent single-species non-spatial systems is most conveniently modelled by non-linear difference equation models, which tend to admit strange attractors at certain domains of their parameter space. Coupling such potentially chaotic systems through linear dispersal terms considerably reduces the chaotic domain in all cases both on the patch level and in terms of the average behaviour of the coupled systems. Threshold-driven, non-linear migration terms can produce dispersal avalanches in coupled map lattices, which can be fairly realistic models of mass migration.

4 Spatially implicit patch models: metapopulations and aggregated interactions

4.1 INTRODUCTION

A recurrent conclusion of the previous chapters is that even moderately realistic multispecies reaction–diffusion models and patchy environment models tend to become analytically intractable with increasing dimensionality. The predictions of spatially explicit mass-interaction models are in most cases also far too abstract for a direct comparison to field data to be possible; any reliable field testing would require the measurement of far too many hard-to-measure parameters. These difficulties explain the relative scarcity of published empirical work aimed at the explanation of experimental or field data with such models.

It was at least in part the need for testable predictions that gave rise to a relatively new class of modelling approach, the result of an obviously fruitful compromise between two equally important, but sometimes also conflicting claims – one for biological interpretability, the other for formal tractability. The compromise regards the way spatial structure is implemented into the formal apparatus of the models of this class: they treat space in an implicit manner, in the sense that locality does not appear in the differential or difference equations formally, neither as spatial coordinates (like in diffusion models), nor as habitat patch indices (like in patch-abundance models).

The class of spatially implicit models includes two broad categories. One is that of **metapopulation models** or **patch-occupancy** models, a special case of which is the basic mainland–island model of island biogeography. The other category includes models of **aggregated parasitoid–host** and **predator–prey interactions** as modifications of the classical Nicholson–Bailey model. Both types incorporate the spatial heterogeneity of the environment somehow – the methods are radically different in the two – but neither keeps track of the topographic information specifying the actual spatial pattern of the populations. A large body of literature is devoted to each of these modelling approaches, which is only in part the result of their attractive simplicity – the other reason is that they yield testable predictions in many

112   Spatially implicit patch models

cases. This is illustrated by two excellent books, one on metapopulation dynamics (Gilpin and Hanski, 1991), the other on aggregated interactions of invertebrate populations (Hassell, 1978), which can both afford to give equal weight to the theoretical and the empirical aspects of their topics. This is a rather infrequent possibility in population dynamics in general, and diffusion–reaction or patch-abundance modelling in particular, unfortunately.

4.2   METAPOPULATIONS AND METACOMMUNITIES

The logic of the simplest patch-occupancy models is perhaps best understood by considering the continuous-time patch models (3.1) and (3.2) of the previous chapter as a reference. If the number of patches is large, and they are similar in size and environmental conditions, a reasonable simplification would be to reduce the representation of within-patch dynamics – which are similar anyway – as much as possible. To do so, one can reduce the relevant population abundance states to presence and absence, and ignore within-patch events other than colonization of empty patches and extinction from occupied ones. Then one can ask about the numbers (or proportions) of empty and occupied patches as a function of time. The diversifying class of patch occupancy or metapopulation models has been built on this simplification in the first place. The approach has been developed later to incorporate details on the local dynamics within patches and the structure of the habitat set, but **not** in a spatially explicit manner.

The word 'metapopulation' denotes a set of local populations that are coupled dynamically through mutual colonizations and recolonizations. I note here that – adhering to the original usage of the concept by Levins (1966, 1969, 1970) for sake of a helpful distinction – I use the term 'metapopulation model' as the synonym of 'patch-occupancy model' in the sequel. By so doing, I deliberately go somewhat against the most recent general trend of the inflation of metapopulation terminology, which also tends to 'swallow' spatially explicit patch-abundance approaches (Taylor, 1990; Hanski, 1991a).

Besides the difference in the state variables of patch-abundance and patch-occupancy models, there are two other serious reasons why they should be treated under separate headings. One of these relates to the difference in the assumptions concerning the spatiotemporal pattern of the environment. In patch-abundance models, the environment within the patchwork is usually assumed constant in time, and in many cases also in space. Neither spatial nor temporal homogeneity is inevitable, of course; patches might differ in their local parameters like the carrying capacities, the growth rates and the interaction parameters of the subpopulations, and the dynamics can be driven by temporal changes in the environment. In metapopulation models, the heterogeneity of the environment both in space and in time is inherent in the assumption of the random occurrence of local extinctions; thus it is a core feature of the approach. The other reason – closely tied to the previous one – for the separate treatment of patch-abundance and patch-occupancy models is the difference in their assumptions on local dynamics. Patch-occupancy models define non-equilibrium dynamics on the local scale, assuming recurrent local cycles of externally driven extinctions (local catastrophes) and recolonizations on the patches, unlike most patch-abundance models (Levin and Paine (1974) and Paine and Levin (1981) are two of the few exceptions), in which local extinction, if it occurs at all in finite time, is a consequence of the internal dynamics (for example, demographic stochasticity) of the system (cf. Verboom et al., 1991; Adler and Neuernberger, 1994).

The metapopulation concept was originally applied for single-species systems (Levins, 1969), but it has been soon extended to multispecies situations, whereby the two categories of presence and absence for a single species on a patch have been replaced with the 2^s categories of possible presence–absence combinations for s species (local biotas). The state variables of the resulting **metacommunity models** are the **frequencies of species combinations**; also in this case no explicit information on local population abundances is used or is available.

4.2.1 Colonization–extinction equilibrium in the basic model

The prototype model of Levins (1969), which defines the dynamics of a single-species metapopulation on a large set of similar habitat patches, is

$$\frac{dp}{dt} = mp(1-p) - ep. \tag{4.1}$$

The state variable of (4.1) is p, the fraction (or relative frequency) of those habitat patches that are occupied by the species; $0 \le p \le 1$. $1 - p$ is the fraction of empty patches, m is the colonization rate, and e is the extinction rate; both rate constants are positive.

What background assumptions can be extracted from this formulation? First of all, the model is of a continuous state variable (p), meaning in this case that we consider a large number of potential habitat patches, one patch representing an infinitesimal fraction of the complete patchwork 'universe'. The habitat patches are identical in size and quality, since both m and e are patch-independent. The time variable is also continuous, implying that both colonizations and extinctions are locally asynchronous, and very frequent within the habitat patchwork. The instantaneous speed of the colonization of empty patches (the first term on the right-hand side of (4.1)) depends on the instantaneous product of the fractions of acceptor (empty) and donor (occupied) patches. As a consequence, it is also implicitly assumed that any one patch is equally accessible to migrant individuals, no matter which patch emitted them. This translates to the assumption of a very (infinitely) fast movement of migrants within the matrix, compared to the time scale of within-patch dynamics. Extinction is a simple exponential process, meaning that in every time unit there is a given constant chance for an occupied patch to lose all its inhabitants, due to some local catastrophe.

Most of these background assumptions seem to be rather unrealistic in biological terms – in a sense, the simplest metapopulation model suffers from the same weaknesses as the logistic model of non-spatial population growth. The similarity is not surprising, given that (4.1) is mathematically equivalent to the single-species logistic with density-dependent **per capita** birth rate $m(1-p)$ and density-independent death rate e. Births are analogous to colonizations, deaths to extinctions, and per capita rates become **per patch** rates in the metapopulation context. The equilibrium point of (4.1) is

$$\hat{p} = 1 - e/m, \tag{4.2}$$

which is positive if $m > e$, and it is, of course, asymptotically stable: there exists a stable dynamic balance between occupied and empty patches if the colonization rate at low p values exceeds the extinction rate; the whole metapopulation goes extinct otherwise. This is the main message of this oversimplified basic model – starting from here, one can ask how this result is altered if more realistic assumptions on colonization, extinction and within-matrix behaviour of migrants are made.

Even very simple modifications to (4.1) can lead to surprisingly strong and testable predictions. If, for example, m is a decreasing function of the average isolation of patches, and e is a decreasing function of patch area – both these assumptions are obviously plausible – we expect from (4.2) that increasing isolation and/or decreasing patch size in a patchwork habitat drives the species towards regional extinction ($\hat{p}$ approaches 0). It is also clear, then, that increasing isolation can be compensated by increasing patch size in terms of equilibrium patch occupancy $\hat{p}$, and *vice versa* (Hanski, 1991b). This conclusion is in good agreement with that concerning habitat fragmentation effects (Section 3.3.4), obtained by the more detailed patch model (3.18); it is also intuitive and – most important – it seems to be in good accordance with field data of different sorts (Hanski and Ranta, 1983; Hanski, 1986; Lomolino, 1986; Lomolino *et al.*, 1989; Opdam, 1991; Harrison, 1991, 1994; Peltonen and Hanski, 1992).

4.2.2 The ghost of within-patch dynamics returns: the asynchronous age-structured model

The Levins model ignores all details on the mechanism of colonizations and extinctions, assuming that m and e – the only parameters of (4.1) – are constants independent of time, space and the fraction p of occupied patches. Albeit, colonization and extinction events are obviously related to local dynamics in many ways, the handy formulation of the patch-occupancy approach was possible only by the omission of explicit local dynamics. There are different implicit methods to smuggle some aspects of local dynamics back into metapopulation models, however, at the expense of mathematical complications that are still manageable. Such improvements to the Levins model affect the colonization and/or the extinction parameters, by expressing them as functions of the state variable p directly or indirectly.

Hastings and Wolin (1989) discuss a metapopulation model based on the additional assumptions (besides those of (4.1)) that

1. each subpopulation grows towards a saturation size (carrying capacity) monotonically with subpopulation age, that is, with the time elapsed since the last colonization event on the patch, and
2. the chance of extinction for a given subpopulation decreases with size, and thus also with age.

Point 1 is an implicit assumption about the local dynamics on the patches, and point 2 establishes the relation between local dynamics and the extinction parameter. The model is formally analogous to that of a non-spatial continuous-time age-structures density-dependent population (Metz and Diekmann, 1986); the analogy is exactly of the same nature as the one between the Levins model and the logistic equation. The model consists basically of two equations, one defining the dynamics of the age distribution for established subpopulations, the other the rate of new establishments through the colonization of empty patches from occupied ones. The first equation is

$$\frac{\partial s(a, t)}{\partial t} = -\frac{\partial s(a, t)}{\partial a} - e(a)s(a, t), \quad (4.3)$$

where $s(a, t)$ is the fraction of patches of age a at time t (that is, s is the instantaneous density function for the age distribution of the occupied patches), and $e(a)$ is the age-dependent extinction rate. Equation 4.3 simply states that the fraction of age a subpopulations can change by ageing, or by going extinct. The fraction of occupied patches at time t can be then expressed as

$$p(t) = \int_0^\infty s(a, t) da. \quad (4.4)$$

The second equation gives the fraction of newborn (of age 0) subpopulations emerging through successful colonizations of empty habitat patches by propagules coming from occupied patches:

$$s(0, t) = [1 - p(t)] \int_0^\infty M(a) s(a, t) da, \quad (4.5)$$

in which $M(a)$ stands for the contribution of age a subpopulations to the number of propagules; thus the integral term is the size of the whole pool of propagules searching for empty patches. Their chance of success, of course, depends on the fraction of empty patches available for invasion, that is, on $[1 - p(t)]$. In fact (4.5) is the boundary condition for (4.3); the $a = 0$ boundary of the age distribution is a source.

The age distribution (and, if there is an explicit relation defined between the age and the size of a subpopulation, also the size distribution) of the metapopulation can be calculated for the colonization–extinction equilibrium of the model, if there is such. A feasible stationary (time-invariant) age distribution $\hat{s}(a)$ with an equilibrium fraction of occupied patches $0 \leq \hat{p} \leq 1$ exists if the expected lifetime contribution of the 'average' patch to the pool of potential colonizing propagules is large enough to replace itself at least, that is, if

$$\int_0^\infty M(a)l(a)\,da > 1, \qquad (4.6)$$

where

$$l(a) = \exp\left(-\int_0^a e(\xi)\,d\xi\right) \qquad (4.7)$$

is the probability that the subpopulation of a newly colonized patch will survive until age a (ξ stands for age).

It can be shown that the density function of the stationary distribution can be expressed in the form

$$\hat{s}(a) = Cl(a), \qquad (4.8)$$

where C is a constant; its actual value depends on the parameters of the model, including those of the colonization function $M(a)$. It is obvious from (4.8) and (4.7) that the shape of the stationary distribution is determined by the age-dependent extinction rate $e(a)$ alone. For any positive $e(a)$ function chosen, the stationary density function is monotonically decreasing with age (since the exponent of (4.7) takes larger negative values with a increasing), which means that the majority of the local populations are always young, that is, small. This is obvious from Figure 4.1, which is an example for $l(a)$, and thus for the shape of the stationary density function $\hat{s}(a)$, representing a special case when the local extinction rate decreases exponentially from $d + b$ to b with subpopulation age: $e(a) = d \exp(-a) + b$. It is assumed that $1 \geq d > b \geq 0$; b corresponds to the extinction rate of the largest possible subpopulation (of a size equal to the carrying capacity of the patch). Hastings and Wolin (1989) show that the equilibrium occupancy satisfying (4.60) is also unique and locally stable, like that of the Levins model.

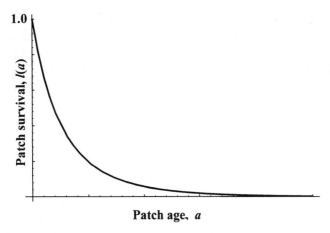

Figure 4.1 The probability $l(a)$ that a patch remains populated until age a; the extinction rate is assumed to decrease exponentially with the age of the local population: $e(a) = d \exp(-a) + b$.

Comparing this model to (4.1), one can see that by introducing plausible assumptions regarding the inner dynamics of the subpopulations, the consequent changes in the colonization and the extinction terms do not alter the qualitative behaviour of the metapopulation, provided the basic assumption of asynchronous recolonizations and local extinctions remains. The model admits a single stable equilibrium for patch occupancy, if its parameters allow for the existence of a positive fixed point. As can be expected, by eliminating the size (that is, age) dependence of the colonization and the extinction parameters M and e from the model, it returns the prototype system in the form

$$\frac{dp}{dt} = Mp(1-p) - ep. \qquad (4.9)$$

4.2.3 Metapopulations with synchronous local dynamics

In the Hastings–Wolin model, not only are colonization and extinction events asynchronous, but also the internal dynamics of the subpopulations are assumed to be out of phase – the age structure of the metapopulation is non-uniform. Hanski (1991b) postulated **synchronous** logistic dynamics within the patches, which assumption is based on the fact that colonization and extinction events are rare on the time scale of local dynamics. He defined an explicit dynamics for the migrants within the pool of migrant propagules, the size I of which depends on the actual (and always uniform!) size N of the local subpopulations, through continuous emigration of individuals from the patches into the pool. The synchrony of local dynamics enters the model through the postulated spatial uniformity of the abundance states of the patches. The rate of change for patch occupancy p depends in turn on the propagule pool size. The model constructed under these assumptions consists of three coupled ODEs, one for patch occupancy, one for local dynamics and one for the propagule pool:

$$\frac{dp}{dt} = i\beta I(1-p) - ep$$
$$\frac{dN}{dt} = rN(1-N) - mN + iI. \qquad (4.10)$$
$$\frac{dI}{dt} = mpN - vI - iI$$

The parameters of this model (each non-negative) are interpreted as follows: r is the growth rate of the subpopulations, scaled so that the carrying capacity of the patches is one; m is the rate of emigration from the patches into the propagule pool; i is the rate of immigration from the pool onto the patches; v is the death rate of the migrants while they are in the pool; β is the probability that an immigration event leads to successful colonization; and e is the local extinction rate as before.

Equation (4.10) can have up to four equilibria, the number depending on the parameters. The interior fixed points, if they exist, are at the intersections of the graphs of two equations relating $\hat{N}$, the equilibrium subpopulation size, and $\hat{p}$, the equilibrium fraction of occupied patches. These equations obtain from the equilibrium condition $dp/dt = dN/dt = dI/dt = 0$ applied to (4.10), which yields

$$\hat{N} = (1-a) + ab\hat{p}$$
$$\hat{N} = \frac{c}{b(1-\hat{p})} \qquad (4.11)$$

$a = m/r$, $b = i/(i+v)$ and $c = e/(\beta m)$ are positive constants. The system admits one interior equilibrium within the $0 < \hat{p} < 1$ interval if $c/b < (1-a)$, and two or none if $(1-a) < c/b$. In the former case, one of the equilibria is stable, the other unstable (Figure 4.2b). If there is only one interior equilibrium, it is either stable, or it is a saddle point (Figure 4.2a and c). There are always two equilibria on the border of the state space. One is the origin, which is always unstable, since the equation for local abundance N in (4.10) can have a positive right-hand side even if I and p are both zero. The other fixed point on the border is – for the same reason – a degenerate state with no occupied patches and an empty pool of propagules ($\hat{p} = 0$ and $\hat{I} = 0$), but still with a positive local density $\hat{N} = 1 - a$, that is, at $(0, 1 - a, 0)$. This equilibrium is rather hard to interpret in biological terms, as it represents a state with no patch to realize the non-zero local density. It is therefore best to consider this fixed point as a state that is never attained in fact, but corresponds to

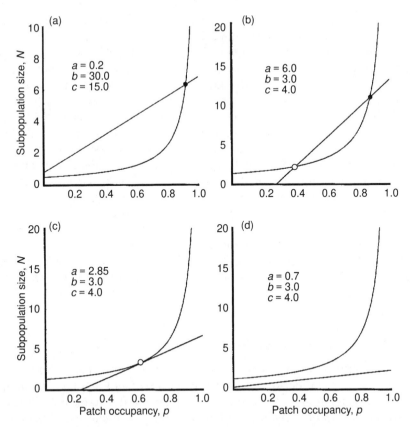

Figure 4.2 Possible interior equilibria of (4.10) with (a) $c/b < (1-a)$: one stable interior fixed point; (b) $c/b > (1-a)$: one stable and one unstable interior fixed point; (c) $c/b > (1-a)$: one interior saddle point; (d) $c/b > (1-a)$: no interior fixed point.

metapopulation extinction in the limit. It is unstable against perturbations into the interior of the state space if $c/b < (1-a)$ (that is, if there is one interior equilibrium) and stable otherwise. If there are two interior fixed points, one of them is always unstable, the other can be stable, and also the degenerate equilibrium is stable. In terms of patch occupancy, this case corresponds to a bistable system, producing either extinction ($\hat{p} = 0$) or persistence ($\hat{p} > 0$) on the metapopulation level, the outcome depending on the initial conditions.

The reason of discussing this model in some detail, in spite of its odd mathematical features arising from the enforced synchrony of local dynamics it assumes, is the possibility of comparing its result to those of the Levins and the Hastings–Wolin models. The conclusions are in qualitative agreement in that if extinctions are frequent (c is large) and local growth is slow

(a is large), there is no interior equilibrium with $\hat{p} > 0$, that is, the whole metapopulation dies out. In the opposite case with rare extinctions and fast local growth, the metapopulation stably persists, which is again in agreement with the conclusion that all previous metapopulation models yield. The difference shows up in the intermediate case, when the system is bistable, that is, when the metapopulation can be either common (of high equilibrium occupancy) or it goes extinct altogether. It seems possible to interpret this difference as a result of just the enforced synchrony built into (4.10). One fact supporting this interpretation is that the asynchrony of local biotic states is correlated to, and to a great extent it is also the ultimate cause of, regional stability in models of spatially explicit patch dynamics, as demonstrated in Section 3.4.3. In the present case, by substituting the asynchrony assumption of local dynamics with the synchronous mechanism in (4.10), the metapopulation becomes less stable in a sense, with an unstable interior fixed point emerging besides the stable one if the extinction rate is high, emigration is intensive and local population growth is fast. Local synchronization may do most harm to a metapopulation system under these conditions, the fate of the metapopulation depending on the initial state of the system.

4.2.4 Rescue effect due to spatial heterogeneity

Hanski (1991b) notes that (4.10) predicts the common field experience of a positive relationship between local abundance ($\hat{N}$) and regional commonness ($\hat{p}$). This relation is obtained by plotting the interior stable equilibria of (4.10) on an $\hat{N}$–$\hat{p}$ pase plane for many different parameter combinations, which results in an elongated cloud of points showing a positive correlation between $\hat{N}$ and $\hat{p}$. Now making the additional – rather straightforward – assumption that the extinction rate decreases with increasing $\hat{N}$, and thus also with increasing $\hat{p}$, one can embed this relation, called the **rescue effect**, into the Levins model as a postulate, without using N as a variable explicitly. A simple and reasonable way of doing so is to assume that the extinction rate is a negative exponential function of patch occupancy, so that (4.1) is modified to

$$\frac{dp}{dt} = mp(1-p) - e_0 \exp(\delta - ap)p, \qquad (4.12)$$

where e_0 and a are the extinction parameters, both positive. Plotting the per patch rates of colonization and extinction as functions of p, the equilibrium occupancies and their stability properties can be determined graphically. There are three possible configurations for the graphs of the **per patch** colonization and extinction rates (Figure 4.3), resulting in zero, one or two interior equilibria, besides the trivial $\hat{p} = 0$. If the graphs of $m(1-p)$ and $e_0 \exp(-ap)$ do not intersect, the only equilibrium state of the metapopulation is the trivial one (complete extinction). If there is one interior fixed point, it is always semistable (that is, stable from above, but unstable from below the interior equilibrium) and the origin is stable. The prospects of metapopulation persistence are bad also in this borderline case, since an infinitesimal perturbation leading to a decrease of p below $\hat{p}$ is fatal to the metapopulation. If two interior equilibria exist, the one lying closer to the stable origin is unstable, whereas the other is stable: the future of the metapopulation is determined by its initial occupancy state. The unstable fixed point separates the basins of attraction of the two stable ones, so that all initial occupancies below the unstable fixed point lead to the extinction of the metapopulation, but if the initial occupancy was high enough, the metapopulation is persistent. By decreasing the extinction parameter e_0 in (4.12), starting with a high value and keeping all other parameters constant, the system passes from the domain of unconditional metapopulation extinction to the interval of conditional persistence (bistability). Comparing this to the original Levins model, we can say that – paradoxically – the rescue effect is destabilizing in this specific form of the dependence of the extinction rate on p. The scenario might change for different shapes of the extinction rate function however, as the destabilizing effect is mainly due to the fact that the extinction rate is very high at low occupancies, so that low occupancy leads to inevitable metapopulation extinction.

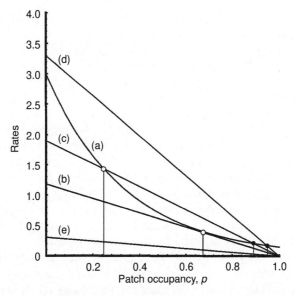

Figure 4.3 Per patch extinction and colonization rates of (4.12): (a) extinction rate with $e_0 = 3.0$ and $a = 3.0$ and (b)–(e) colonization rates with (b) $m = 3.3$: one stable interior fixed point; (c) $m = 1.9$: one stable, one unstable interior fixed point; (d) $m = 1.18$: one 'semistable' interior fixed point; and (e) $m = 0.3$: no interior fixed point.

4.2.5 Patch size and quality effects: phenomenological model

The size and the environmental quality of a habitat patch obviously affect both the probability of its successful colonization when empty and the risk that the resident subpopulation will go extinct when occupied. To take these habitat differences into account in a modified Levins model, one can assume a relationship between patch occupancy and the average environmental quality of the fraction of actually occupied patches, and express the colonization and extinction rates as functions of that quality. It is reasonable to assume that extinction slows down as patch occupancy decreases, because the poorer quality patches are the first to become extinct, and thus the average quality of the remaining populated patches increases. This is reflected in a decrease of the **per patch** extinction rate as p decreases, which means that the extinction rate is an increasing function of p at low occupancies. Note here that this assumption states that patch quality and patch occupancy are canonical variables: when only a few patches are occupied, these must be the best ones – all other situations are impossible in this model.

Hanski (1991b) takes the patch quality effect, together with the rescue effect, into account by multiplying the extinction rate in (4.12) with p^τ:

$$\frac{dp}{dt} = mp(1-p) - e_0 \exp(-ap) p^{1+\tau}. \tag{4.13}$$

τ is positive; the smaller it is, the larger the difference among the patches in habitat quality, and the fewer the safe (large or good quality) patches. Extremely small values of τ correspond to situations close to the mainland–island case with a few safe patches and many vulnerable ones.

The habitat heterogeneity effect can counterbalance the rescue effect, so that the **per patch** extinction rate increases for small p, thus allowing for the possibility of up to three interior equilibria of (4.13) (Figure 4.4). The trivial equilibrium $\hat{p} = 0$ is always unstable, except for the unimportant case of no colonization, that is, $m = 0$. Let us see what happens to the system if we increase the colonization parameter m, while keeping all other parameters constant! For small values of m, there is one interior fixed point, which is always stable, but it stays close to the origin, implying a high risk of global extinction. Having passed a critical level of dispersion (which implies a semistable equilibrium), two additional fixed points appear, one unstable and the other stable. Now there are three interior equilibria, of which two are stable (the one closest to, and the one furthest from, the origin); the third between them is unstable. The system is bistable also in this case; the faith of the metapopulation depends on the initial conditions: either it approaches the low-occupancy equilibrium state, or it attains the high one. For sufficiently high dispersal, the high-occupancy interior equilibrium becomes unique, implying a safe level of presence for the metapopulation within the habitat patchwork, irrespective of initial conditions.

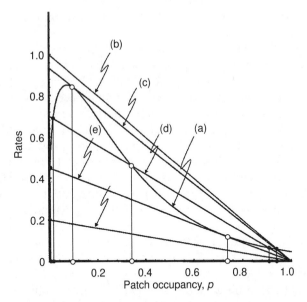

Figure 4.4 Per patch extinction and colonization rates of (4.13): (a) extinction rate with $e_0 = 2.5$, $a = 4.0$ and $\tau = 0.3$ and (b)–(f) colonization rates with (b) $m = 1.0$: one stable interior fixed point; (c) $m = 0.93$: one semistable and one stable interior fixed point; (d) $m = 0.7$: two stable and one unstable interior fixed point; (e) $m = 0.45$: one stable and one semistable interior fixed point; (f) $m = 0.2$: one stable interior fixed point.

This model also admits bistability, as does (4.12), but now with the origin (metapopulation extinction state) always unstable. The main prediction for this situation would be that if a certain number of similar, but essentially non-interacting populations follow (4.13), a bimodal distribution of equilibrium patch occupancy should be observed with many species being either rare or common in the patchwork habitat. This prediction of (4.13) is the so-called **core-satellite species hypothesis**, which is relatively easy to test on the field – in fact it turned out to be applicable in quite a few documented field situations (cf. Hanski (1991b) and references therein).

The bimodal occupancy pattern is maintained by the habitat heterogeneity effect and the rescue effect in such a way that the low-occupancy stable state is dominated by the former, whereas the high-occupancy stable state is dominated by the latter. Species within the bistable parameter regime can realize either of these, depending on the history of the metapopulation and the habitat, but others – those with a single interior equilibrium – will occupy either a few or many patches, irrespective of the initial state of the metapopulation. Core species owe their regional commonness to the rescue effect; satellite species are maintained by the habitat heterogeneity effect.

4.2.6 Patch size and quality effects: mechanistic model

Many of the extensions of the Levins model so far considered are included in the most detailed single-species metapopulation system constructed to date, due to Gyllenberg and Hanski (1992) and Hanski and Gyllenberg (1993). This model accounts for many different aspects of within-patch and within-matrix dynamics through concrete mechanisms, albeit still in a spatially implicit manner. I will not show the model in detail (the reader is referred to the original papers), but give a short – and thus, inevitably, somewhat sloppy – description of its assumptions and conclusions instead. The model is very close, both in spirit and in its formal structure, to the Hastings–Wolin type age-structured partial differential equation models (4.3) and (4.5) discussed in Section 4.2.2, with some essential modifications. The most significant of these is that the **size structure** of the patchwork is an explicit input of the model, given as a distribution of local carrying capacities, unlike in the Hastings–Wolin model, which assumes environmentally identical patches, all with the same carrying capacity. Therefore, the age distribution $p(t, a)$ of the metapopulation is not an unambiguous descriptor of its actual state any more – a complete specification is given by the distribution of subpopulation sizes (x) on the distribution of patch sizes (that is, carrying capacities, y) at any time t, that is, the state variable of the model is $p(t, x, y)$. The dynamics within the pool of migrants is explicitly modelled; emigration, immigration and the colonization of empty patches might all depend on x and y. It is important that immigration affects the dynamics on occupied patches, which is a feature of the model in common with the discrete state system (4.14) below.

The model, in its most general form, is very difficult to study analytically. Hanski and Gyllenberg (1993) decided to address rather specific problems, which allow for substantial simplifications in many respects, but even with these, the appropriate tool of investigating the dynamics of the system is numerical analysis mostly. The main simplification is that the size distribution of the patches is assumed discrete, local carrying capacities taking only two possible values, y_1, 'small' and y_2, 'large', in proportions q and $(1 - q)$, respectively. By adjusting y_1/y_2 and q, it is possible to approximate different habitat structures, from an archipelago of similar size patches to a mainland–island arrangement with a few large patches and many small ones. One can also adjust the migration, colonization and extinction terms to fit the model to specific situations. The mainland–island case can be, for example, characterized by zero extinction rate on large patches (mainlands), and negligible migration from small patches to large ones, and optionally also among small patches. The rescue effect can be generated by a sharply decreasing dependence function of the extinction rate on local population size.

All numerically studied parameter combinations of the model produced one to four equilibria in terms of patch occupancy, the set of fixed points always including the zero occupancy (metapopulation extinction) state. Just as in the phenomenological model (4.13), the local stability pattern is fully specified by the number of existing fixed points. If there is only one, it is the zero state, and it is locally stable. If there are two equilibria, the zero state is unstable, and the interior fixed point is stable. The existence of three equilibria corresponds to the bistable case with conditional metapopulation extinction. Four equilibria imply also bistability, but now the zero state is unstable, and two of the interior equilibria are stable, separated by a third unstable one – note that this is also the configuration of (4.13) at the bistable parameter regime. Thus the emergence of an unconditionally persistent bistable case also predicts a bimodal species distribution, which fits well to the core-satellite species hypothesis and related field data.

Although the Gyllenberg–Hanski model is similar to (4.13) in the qualitative outcomes it produces, it is also different in some respects. The most obvious difference comes directly from the fact that (4.13) is a phenomenological model: it includes both the rescue effect and patch size or quality heterogeneity effect as basically ad hoc functions of patch occupancy, with no actual mechanism specified. The Gyllenberg–Hanski model is mechanistic, postulating on details of the size structure of the patchwork, within-patch dynamics, and colonization–extinction events in any realization. The results are similar to those of the phenomenological model as long as the concrete mechanisms specifying the patch size (or quality) effect and the rescue effect produce qualitatively similar dependences of per patch colonization and extinction rates on patch occupancy. This expectation is reinforced by the numerical results of Hanski and Gyllenberg (1993), indicating that the more extreme differences in the distribution of carrying capacity and extinction probability one assumes between small and large patches, the more likely the appearance of a bimodal equilibrium distribution becomes, both in the archipelago and in the mainland–island situations.

4.2.7 Multistate metapopulation models

Besides implicit assumptions on the dependence of colonization and/or extinction rates on the actual patch-occupancy level or on other parameters of the model, there is another way to smuggle some local dynamics back into patch-occupancy models, namely by setting up more than one category of occupied patches, according to a distinction by local abundance. The simplest possible extension of the Levins model in this direction requires splitting up the 'occupied' state to 'occupied at low local density' and 'occupied at high local density' categories. Thus, a patch can be empty, it might carry a small subpopulation or a large one: we have three possible states instead of two. The differential equations of the model thus constructed specify the rates of change for the relative frequencies p_e, p_s and p_l of the empty, the small subpopulation and the large subpopulation states, respectively:

$$\frac{dp_e}{dt} = ep_s - mp_l p_e$$
$$\frac{dp_s}{dt} = mp_l p_e + cp_l - \lambda p_s - ep_s - fp_l p_s, \qquad (4.14)$$
$$\frac{dp_l}{dt} = \lambda p_s + fp_l p_s - cp_l$$

where e is the extinction rate of small subpopulations, m is the colonization rate of empty patches by propagules coming from large subpopulations, c is the transition rate of large subpopulations into small ones by 'partial extinction', λ is the transition rate of small subpopulations into large ones (corresponding to local population growth), and f is the rate of small subpopulations becoming large ones due to immigration of propagules from large subpopulations (Hanski, 1985). All these parameters are non-negative. Note that large subpopulations cannot become extinct immediately, only by passing through the small density state. Since the values of p are relative frequencies of the possible states of patches, trajectories are confined to the two-dimensional standard simplex satisfying

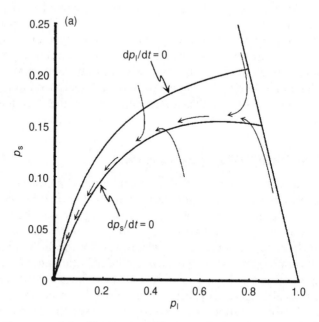

Figure 4.5 Zero growth isoclines and equilibria of (4.14): (a) $e = 0.2, m = 0.1, \lambda = 0.1, f = 0.5, c = 0.13$: no interior fixed point, metapopulation extinction; (b) $e = 0.2, m = 0.4, \lambda = 0.05, f = 1.0, c = 0.2$: one stable interior fixed point; and (c) $e = 0.2, m = 0.4, \lambda = 0.05, f = 0.3, c = 0.04$: one stable and one unstable interior fixed point.

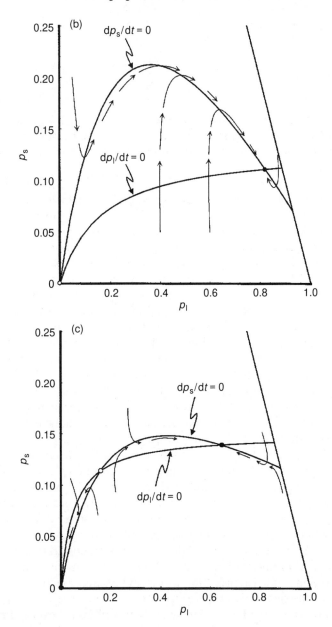

Figure 4.5 Continued

$$p_e + p_s + p_l = 1, \tag{4.15}$$

and thus also

$$\frac{dp_e}{dt} + \frac{dp_s}{dt} + \frac{dp_l}{dt} = 0 \tag{4.16}$$

holds, expressing the trivial fact that the relative frequency of any one state can grow only at the expense of the other two. This constraint makes one of the three equations in (4.14) redundant – any two are sufficient to specify the dynamics.

Hanski (1985) studied the stability properties of the equilibria of (4.14) within the p_s–p_l phase plane, the origin of which corresponds to metapopulation extinction with $p_e = 1$. This is always a fixed point of the system, but not always stable. In fact, (4.14) can admit one to three equilibria, depending on the parameters. If there is only one equilibrium (Figure 4.5a), it is the origin which is globally stable, so that the whole metapopulation dies out in the long run. As one can expect by intuition, this happens if the extinction and the density reduction rate (e and c) are large compared to the colonization rate m and the density growth rate λ. Better chances of local persistence (smaller e and c) and more intensive dispersion (larger m) first generate two interior fixed points besides the still stable origin, but only if large subpopulations contribute significantly to the increase of small ones (that is, if f is large enough; Figure 4.5b). One of the emerging interior fixed points is stable, the other (the one closer to the origin) is unstable, so that trajectories converge towards either metapopulation extinction or regional persistence, depending on where they started from (bistable case). By reducing e and c, or by increasing m further, the unstable interior fixed point disappears and the origin loses stability, leaving the previously locally stable interior equilibrium globally stable (this property corresponds to unconditional metapopulation persistence; Figure 4.5c). A large enough subpopulation growth rate (λ) has the same effect, which is also to be expected by intuition. The parameter domains of these cases are also plotted on the stability map in Figure 4.6.

Comparing this result to that of the age (size) structured Hastings–Wolin model, which predicts either unconditional metapopulation extinction or unconditional persistence, but not bistability, one might wonder about the cause of the qualitative difference between them. The question is pertinent because these models are similar in terms of their basic assumptions, with the exception of three important points. One is that the Hastings–Wolin model assumes a continuum of local population size categories, instead of the three discrete ones in (4.14). The second is that (4.14) allows only for density reduction on dense patches, instead of complete extinction. The third difference regards the positive effect of migrants on local population growth ($f > 0$); in the continuous model, no such effect is assumed at all.

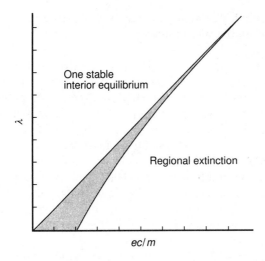

Figure 4.6 Stability map of (4.14), for $c/f = 0.1$. The shaded region of the parameter space corresponds to alternative stable states (that is, bistability) (after Hanski (1985)).

Hanski (1985) and Hastings (1991b) both show that (4.14) loses the bistability property if f, the interpatch migration parameter, is small. This seems to support the conclusion that it is the third difference, that is, the positive effect of migrants on local dynamics that is responsible for the emergence of an unstable interior fixed point, and thus bistability in (4.14). Such a conclusion comes even closer with a modification of the Hastings–Wolin model, due to Hastings (1991), who showed that the second difference is probably **not** responsible for bistability. His model also included what he calls 'partial disasters' – catastrophic events that nonetheless do not devastate the patches completely and only reduce the size of the resident subpopulations – in a Hastings–Wolin type system. The modified model has also been shown to admit a unique interior equilibrium, like the original. In other words, bistability behaviour does not show up in the continuous size-structured model with partial disasters. The effect of the first difference, that is, a finite number of discrete states versus a continuum of abundance states, seems not to be relevant from the viewpoint of bistability, since neither the Hastings–Wolin model, nor (4.14) can produce it without the positive effect of migration on local dynamics. It would be a possible – but obviously also very tedious – check of this conjecture to see whether a model of the (4.14) type with more than three density categories and migration affecting local dynamics, would yield different conclusions.

4.2.8 Interacting metapopulations: the structure of metacommunity models

The explanation of the core-satellite species distribution pattern (Sections 4.2.5 and 4.2.6) rests on two important background assumptions, which connect the single-species metapopulation approach to multispecies situations. First, many species are assumed to follow the same colonization–extinction dynamics on the habitat patchwork, one differing from the other only in parameter values. Second, the metapopulations are independent of each other: no demographic or dispersal interactions are taken into account. Even if the first assumption is acceptable in many situations, the second applies only to rather specific field cases; the chance of occurring decreases with the number of species involved. The interaction of metapopulations cannot be omitted in most real situations; this fact calls for a more general theoretical approach.

Thinking in terms of patch occupancy, the single-species model can be naturally extended to include multispecies systems, in which metapopulations not only co-occur independently on the same set of habitat patches, but they also interact. In the single-species case, the patches can be in one of two possible occupancy states: empty or occupied. With two species, each can be present or absent on a patch, and the number of possible patch-occupancy states is four, possible local biotas including none, any one or both species. In the general case with a species pool of size s, a patch can be in any one of the 2^s potential occupancy categories or local biota types.

A local biota can change by the extinction of previously present species from the given habitat patch, and by the immigration of previously absent species from other patches. The rates of such biota transformations can be assumed to depend on the numbers (or proportions) of patches in each possible local biota type. This is the general idea behind any metacommunity model, independently of the actual type (or types) of population interactions considered. Both demographic (within-patch) and dispersal (within-matrix) interactions are implicit in the rates of biota transformations.

Metacommunity models can be implemented as sets of non-linear differential or difference equations (e.g., Horn and MacArthur, 1972; Slatkin, 1974, Hanski, 1983; Caswell and Cohen, 1991a, b; Nee and May, 1992), one equation specifying the rate of change in the fraction of patches in each biotic state. The methodological limitations of metacommunity modelling come from this fact, since by increasing the number of species considered, the number of non-linear differential or difference equations in the corresponding model increases to $2^s - 1$, whereas the number of parameters becomes of the order of 2^{2s}. This leads to analytical and also numerical intractability very soon, for exactly the same reasons as does increasing the number of species in a non-spatial Lotka–Volterra system or a patch-abundance model. Metacommunity modelling is therefore also an oligospecies approach in essence. It is very effective within this realm, because it allows for judging the effects of many different assumptions regarding local dynamics, dispersion and habitat heterogeneity in a patchwork habitat by introducing tractable modifications, unlike in patch–abundance models which can incorporate such complications only at the expense of a significant increase in the number of model parameters, and thus an inevitable loss of interpretability.

4.2.9 Competitive metacommunities: continuous time models

The simplest possible modification of the Levins model to incorporate competitive interaction in a two-species metapopulation model is due to Slatkin (1974), who set up the following system:

$$\frac{dp_0}{dt} = -(m_1 x_1 + m_2 x_2) p_0 + e_1 p_1 + e_2 p_2$$
$$\frac{dp_1}{dt} = m_1 x_1 p_0 - [e_1 + (m_2 - \mu_2) x_2] p_1 + (e_2 + \varepsilon_2) p_3$$
$$\frac{dp_2}{dt} = m_2 x_2 p_0 - [e_2 + (m_1 - \mu_1) x_1] p_2 + (e_1 + \varepsilon_1) p_3$$
$$\frac{dp_3}{dt} = (m_1 - \mu_1) x_1 p_2 + (m_2 - \mu_2) x_2 p_1 - (e_1 + \varepsilon_1 + e_2 + \varepsilon_2) p_3$$

(4.17)

The possible state transitions within this system are illustrated graphically in Figure 4.7. p_0 is the fraction of empty patches, p_1 and p_2 correspond to those patches populated by species 1 and 2, respectively, and p_3 is the fraction of patches with both species present. $x_i = p_i + p_3$ denotes the fraction of patches occupied by species i. Since the values of p are fractions, one of the differential equations of (4.17) is redundant and can be replaced by the obvious relation $p_0 + p_1 + p_2 + p_3 = 1$. m_i is the constant rate of colonization of empty patches by species i, which is reduced by μ_i if the other species is already present on the patch to be colonized; e_i is the extinction rate on patches with only species i present, which is increased by ε_i if the patch is occupied by both species; μ_i and ε_i are the interspecific interaction parameters of the model, determining the measure of the decrease in colonization chance and that of the increase in extinction risk caused by the presence of the competitor species. Note that if these parameters are set to 0, the species become disconnected dynamically, and (4.17) reduces to the Levins model with two independent metapopulations incidentally co-occurring on the same habitat patchwork. This can be shown by adding dp_1/dt and dp_3/dt or dp_2/dt and dp_3/dt and setting $\mu = \varepsilon = 0$, which results in two independent Levins models for x_1 and x_2, respectively.

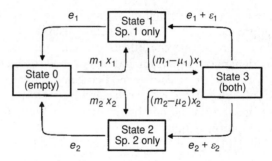

Figure 4.7 Possible transition routes and the corresponding per patch rates for (4.17). Note that double transitions (state 0 → state 3; state 1 → state 2 and their reverses) are not allowed.

The interesting question to ask now regards the possibility of the coexistence of the two species, provided both would be persistent without the other one present, that is, if $e_i/m_i < 1$ (note that this is the persistence condition of the Levins model). The straightforward approach would be to determine the interior equilibria of the system, if there are such, and study their stability properties. Unfortunately this is not at all easy in this case. Although (4.17) might seem rather innocent at first sight, it turns out to be very complicated algebraically, if one tries to calculate its interior equilibria. What is easy to calculate are the two boundary equilibria Q_1 and Q_2, corresponding to the cases with one of the metapopulations missing from the system:

Q_1: $\hat{p}_0 = e_1/m_1$, $\hat{p}_1 = 1 - e_1/m_1$, $\hat{p}_2 = \hat{p}_3 = 0$, and

Q_2: $\hat{p}_0 = e_2/m_2$, $\hat{p}_2 = 1 - e_2/m_2$, $\hat{p}_1 = \hat{p}_3 = 0$.

Performing local stability analysis on these fixed points, the possible combinations of outcomes are

1. both Q_1 and Q_2 are unstable;
2. Q_1 is stable, Q_2 is unstable;
3. Q_2 is stable, Q_1 is unstable;
4. both Q_1 and Q_2 are stable.

Case 1 means that both species can invade the patchwork if the other is present, that is, coexistence is inevitable (recall that the species are persistent if left alone), albeit the properties of the limit set are not known in detail, except that it is within the positive quadrant of the state space. Numerical analysis of the system suggests, however, that the limit set is a stable fixed point. One expects species 1 to exclude species 2 in case 2 and the reverse to happen in case 3, where the expectation seems to be justified by numerical analysis. Note that the behaviour of the system shows a striking similarity to that of the two-species competitive Lotka–Volterra model so far (cf. Section 3.3, equation (3.3)). The analogy extends also to case 4, which corresponds to the conditional exclusion of one of the metapopulations from the patchwork habitat. Which one of the species wins depends on the initial occupancies.

The local stability analysis for the most general form of (4.17) again yields very long and complicated formulas, from which it is hard to extract the criteria of coexistence. It is more instructive to study biologically important special cases with certain constraints imposed on the parameter configuration. One such specific case is, for example, to consider two like species with all their corresponding parameters being equal. Slatkin (1974) shows that under this assumption, both boundary equilibria are locally unstable; thus the species are coexistent. It can be demonstrated by numerical solutions of (4.17) with different parameter sets and initial conditions that given the unstable boundary equilibria, the system admits a single stable internal fixed point, corresponding to stable metapopulation coexistence. This result apparently contradicts Gauss's competitive exclusion principle, which would predict the elimination of one of the two similar species. The resolution of the contradiction lies in the existence of two different spatial scales in metapopulation modelling; exclusion on the patch level does not necessarily also mean exclusion on the patchwork level. This is a moral in common with those of spatially explicit models like (3.16) (cf. Section 3.2).

Another simplifying assumption might be setting the extinction rates independent of patch occupancy, that is, postulating $\varepsilon_1 = \varepsilon_2 = 0$. This is the so-called **migration competition case** in the terminology of Levins and Culver (1971), where it is only the chance of patch colonization that is affected by the presence of the competitor – once the patch is colonized, the species behaves as if the competitor was not there (only dispersal interaction). This system allows for coexistence if the species are not very different, or unconditional extinction for one of the species otherwise; more precisely, if $\hat{p}_i > \hat{p}_j$, then species i can always invade Q_j (that is, Q_j is locally unstable), but if both $\hat{p}_i < \hat{p}_j$ and $\mu_i/m_i > \hat{p}_i/\hat{p}_j$ hold, it cannot (that is, Q_j is locally stable). From these conditions it follows that either both boundary equilibria are unstable, implying coexistence (this is the case if, for example, $\hat{p}_1 < \hat{p}_2$, but $\mu_1/m_1 < \hat{p}_1/\hat{p}_2$ – that is, species 1 performs worse if left alone in the habitat patchwork, but suffers less in terms of migration reduction from species 2 in a competitive situation), or one is stable and the other is unstable, implying competitive exclusion.

If competition affects local extinctions only, but not colonizations, that is, $\mu_1 = \mu_2 = 0$ (**extinction competition case**), the result is qualitatively similar: either one of the species excludes the other, or they coexist – no bistability, and therefore no conditional exclusion dependent on the initial occupancy state of the patchwork is possible. Unconditional exclusion occurs under similar assumptions than in the migration competition case, namely if one of the species attains a high equilibrium occupancy if left alone, and it is also a strong competitor of the other species in terms of increasing its extinction rate effectively. Although similar in spirit, the precise conditions look much more complicated and much less instructive than those for the migration competition system, so instead of reproducing them here, the interested reader is referred to Slatkin (1974) for the exact formulae.

4.2.10 Discrete time metacommunity models

Caswell and Cohen (1991b) implemented three different competition mechanisms into a discrete time metapopulation model, and studied the criteria of coexistence in that framework. The model is formulated for s species as a non-linear Markov chain on the relative frequencies x_k ($k = 1, \ldots, 2^s$) of the 2^s possible occupancy states (biotas) on a large set of similar habitat patches. The actual studies involved two or three species. With two species, the four possible biotas are indexed as follows:

$k = 1$: no species present;
$k = 2$: species 1 present, species 2 absent;
$k = 3$: species 1 absent, species 2 present;
$k = 4$: both species present.

The relevant component processes of the model are colonizations, extinctions due to abiotic effects and extinctions due to biotic (that is, competitive or predative) effects. Colonizations occur randomly according to a Poisson process, the parameter of which is proportional to the size of the pool of dispersers, which is in turn proportional to the fraction f_i of the patches occupied by, and the dispersivity d_i of, species i. Under this assumption, the probability that a patch will be encountered by at least one propagule of species i, and thus potentially colonized, is

$$C_i = 1 - \exp(-d_i f_i). \tag{4.18}$$

f_i is the sum of the relative frequencies x_k of all those biotas which include species i. Because of the discrete time steps, simultaneous colonizations by more species are allowed. The complete devastation of an inhabited patch, due to some abiotic disturbance, is of a constant probability p_d, regardless of what the actual biota was on the patch before the catastrophe. For the two-species competition model, the competitive dominance hierarchy is fixed so that species 1 can always exclude species 2, but the reverse event is not possible. Competition enters the model as a constant probability p_c of the local exclusion of species 2 by species 1, if they happen to co-occur on a patch.

The three different patterns of competitive interaction that can be considered within this framework correspond to the tolerance, facilitation and inhibition mechanisms of succession, suggested by Connell and Slatyer (1977). Tolerance competition means that the competitively superior species 1 can always colonize a patch, independently of whether it is occupied by species 2 or not, but species 2 can colonize empty patches only. The facilitation case implies that the subordinated species 2 can colonize empty patches only, whereas species 1 needs the presence of species 2 on a patch to be able to colonize it, but then it can exclude the resident competitively. Inhibition competition occurs if neither species can invade a patch where the other is resident, but both can, even simultaneously, if the patch is empty. Of course species 2 can be excluded later from simultaneously colonized patches. Note that the differences of these three mechanisms regard colonization events only: in fact they are variants of migration competition.

With each of the competition mechanisms above, one can construct a transition matrix $\mathbf{A}_x$ for the state transition of the vector $\mathbf{x}$ of biotal frequencies from time t to $t+1$. For the tolerance competition case with two species, for example, the transition matrix is

$$\mathbf{A}_x = \begin{pmatrix} (1-C_1)(1-C_2) & p_d & p_d & p_d \\ C_1(1-C_2) & 1-p_d & 0 & (1-p_d)p_c \\ (1-C_1)C_2 & 0 & (1-C_1)(1-p_d) & 0 \\ C_1C_2 & 0 & C_1(1-p_d) & (1-p_d)(1-p_c) \end{pmatrix} \quad (4.19)$$

which depends on $\mathbf{x}$ because the colonization parameters C_i are functions of $\mathbf{x}$. It is through these parameters where non-linearity enters the system, making it practically intractable analytically. Caswell and Cohen (1991b) studied this model for all the three competition mechanisms numerically, iterating the matrix equation

$$\mathbf{x}(t+1) = \mathbf{A}_x \cdot \mathbf{x}(t) \quad (4.20)$$

by computer. Scanning the parameter space of the two-species model with many different parameter combinations, they found that trajectories settle on a fixed point in all cases. Without disturbance, that is, with $p_d = 0$, the superior competitor always won regionally, but for intermediate values of p_d, the species tended to coexist irrespective of the competition mechanism, provided the subordinate species was a good enough disperser. The criteria of coexistence are that the fugitive species be a good colonizer, and local catastrophes be frequent enough. This prediction is in good agreement both with intuitive expectations and with the results of many experiments (e.g., Connell, 1978; Huston, 1979; Dethier, 1984) demonstrating the coexistence of competitively dominant species with so-called fugitive ones. Fugitive species are subordinate when it comes to competitive interaction, but they are good dispersers, and therefore they are able to find and colonize empty patches soon after such empty habitats are created by local catastrophes. If the catastrophes are frequent enough, there are always empty patches in sufficient numbers for the fugitive species to occupy, thus for maintaining itself within the region in spite of the presence of the dominant competitor.

By extending the model to three competing species, S_1, S_2 and S_3, the possibility of an intransitive cycle of competitive relations (S_1 beats S_2, S_2 beats S_3, S_3 beats S_1) arises, which can itself ensure the regional coexistence of competitors in the metacommunity by driving asynchronous local cycles of competitive succession on the patches. If the competitive relations are hierarchical (that is, S_1 beats S_2 and S_3, and S_2 beats S_3), disturbance probability and the dispersing abilities of the subordinate species S_2 and S_3 are again important in determining the outcome of competition on the regional scale, just as in the two-species case.

The same model is easily modified to study other types of interspecific interactions by adjusting the parameters C_i and p_c conveniently. Caswell and Cohen (1991b), for example, consider two competing prey populations and a generalist predator feeding on both without preference. They assume that one of the prey species always loses competition if the two happen to co-occur on the same locality, so that the uniform monoculture of the dominant competitor would be the ultimate state of the system without the predator and local catastrophes. If the predator manages to colonize a patch, it eliminates all its resident prey with a probability p_e. The three species turn out to be coexistent for large enough values of p_e, and a sufficiently large dispersion rate for the weaker prey. This is to be expected, by noting that the predator acts as a local catastrophe from the viewpoint of the prey populations, and local catastrophes help the inferior competitor to persist if it is a good colonizer. The only difference between external catastrophes and predation in this model is that the latter is not of a constant probability, but dependent on predator frequency. This is a clear case of the predator-mediated coexistence of otherwise exclusive competitors on a patchwork habitat.

4.2.11 Comparing patch-abundance and patch-occupancy models

Although the fast developing metapopulation and metacommunity topic is not nearly exhausted by the above selection of models, we are now in a position to attempt the comparison of the morals of the metapopulation approach to those of patch-abundance modelling. An interesting conclusion of the comparison is that the predictions these two model types provide are in many cases rather similar from the biological viewpoint, in spite of the obvious differences in assumptions and mathematical structure. Metapopulation systems are often regarded as phenomenological approximations to more detailed patch-abundance models, because the former predicts essentially similar outcomes than the latter, through a much simpler formalism. The similarity seems very obvious, for example, if one compares the Levins model (4.1) to the spatially explicit single-species multipatch model (3.18). Both predict global stability for the unique interior fixed point, if it exists (for the global stability of patch abundance type models, see Freedman and Takeuchi (1989a) and Hofbauer (1990)), and thus stable persistence for the species. It is important to see, however, that the similarity is superficial in this case: the global stability property emerges due to different mechanisms in the two systems. In the single-species patch-abundance model, global stability means stability on both the local and the regional scale, that is, at equilibrium, all the patches are always occupied in a temporally invariant abundance pattern. This implies, of course, also regional stability. The Levins model, and also patch-occupancy models in general, assume that none of the patches is stationary: events of extinction and recolonization alternate on the local scale, and it is only the regional occupancy state of the patchwork that stays constant at the globally stable fixed point. Implicit non-equilibrium local dynamics generate explicit equilibrium dynamics on the regional scale.

On the other hand, both model types confirm the general conclusion that spatial heterogeneity tends to have a stabilizing effect on population dynamics. In a sense, the metapopulation approach is quite similar to oscillatory (e.g., predator–prey) patch-abundance systems, at least in that the ultimate source of stability is the **asynchrony of local dynamics** in both. The difference is that local asynchrony is internally (if at all) generated in oscillatory patch-abundance systems, whereas it is externally enforced (by asynchronous local catastrophes) on metapopulation models.

4.2.12 Connection to island biogeography: incidence function models

Some important aspects of the theory of island biogeography (MacArthur and Wilson, 1967) can be directly related to the metapopulation approach (Hanski, 1991a). In its basic form, the habitat region of interest for island biogeography is an archipelago that consists of many environmentally similar and dynamically unconnected islands, and a mainland of a constant and eternal source capacity that repeatedly 're-inoculates' extinct subpopulations. The islands are unconnected in the sense that the contribution of inter-island migration to the number of colonists is insignificant as compared to that of those coming from the mainland. With this assumption, the Levins model (4.1) can be modified so that the colonization rate is not a function of the fraction p of occupied patches directly. Then the speed of colonization depends only on the fraction of empty patches $1 - p$, so that the modified system becomes

$$\frac{dp}{dt} = m(1-p) - ep. \tag{4.21}$$

This system has always a positive equilibrium at $\hat{p} = m/(m + e)$, which is obviously stable.

138   Spatially implicit patch models

This model would not be very interesting in itself: its significance lies in the possibility it offers for measuring its key parameters in the field, provided the so-called **incidence histogram** of a given species on a given archipelago is known. Supposing that the extinction rate e decreases with the area a of the islands within the archipelago according to $e = e'a^{-\alpha}$ ($0 < e' < 1$ and $\alpha > 0$ are extinction constants), one gets an equilibrium occupancy value by modifying the equilibrium to

$$\hat{p}(a) = \frac{m}{m + e'a^{-\alpha}} = \frac{1}{1 + (e'/m)a^{-\alpha}}. \tag{4.22}$$

for the subset of islands which are of area a. For $a < a_0 = (e')^{1/\alpha}$, e would have values larger than one; therefore it is assumed that $\hat{p}(a < a_0) = 0$: very small islands cannot maintain local populations at all. If m and e are considered as per unit time probabilities of colonization from the mainland and local

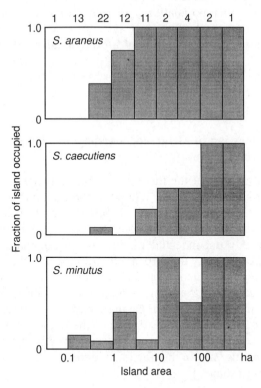

Figure 4.8 Incidence histograms for three species of shrew (*Sorex araneus*, *S. caecutiens* and *S. minutus*) on 68 islands in lakes of Finland. The islands are grouped in logarithmic size classes on the horizontal axis; the vertical axis gives the fraction of occupied islands in the corresponding size class. The number of islands in each size class is given by the figures above the histograms (after Peltonen and Hanski (1992)).

extinction, respectively, and the equilibrium occupancies $\hat{p}(A)$ are given for each size class A, then $\hat{p}(A)$ is an incidence function or incidence histogram (Hanski, 1991b, 1994a, b) of the species on the archipelago or patchwork habitat. The theoretical incidence function specifies for each size class of habitat patches the expected equilibrium occupancy of the species on the islands falling within that size class. Empirical incidence functions can be easily obtained for species within a large enough patchwork (Figure 4.8).

If one can fit (4.22) to such empirical incidence histograms, the parameters e'/m and α of 4.22 can be estimated. If also a_0 is known (the island size at which e has a value of 1), e', and thus also m, can be calculated. In all, it is possible to relate empirical incidence histograms to those predicted by the model, and guess all the colonization and extinction parameters by fitting the two. Then the predicted parameter values can be compared to colonization and extinction rates measured directly and independently in the field, to confirm or to falsify the assumption that it was in fact the hypothesized mechanism that led to the observed incidence function.

Hanski (1994a, b) developed a more realistic incidence function model by extending the patch-occupancy approach to a **spatially semi-explicit** system. Considering the relative spatial relations of a finite set of habitat patches enables the probability of colonization on patch i to be a function of its biotic isolation either from the mainland (mainland–island case), or from the other patches (archipelago case). Specifically, in the single-species model, Hanski (1994a) and Moilanen and Hanski (1995) assume that the 'invasiveness' S_{ij} of patch j to patch i is a negative exponential function of the distance d_{ij} between the two patches, and it is linearly proportional to the area A_j of patch j: $S_{ij} = p_j \exp(-\alpha d_{ij}) A_j$. α is a constant characterizing dispersivity; the smaller it is, the better the species disperses. p_j is also a constant expressing the propensity of the species to emigrate from patch j: the better environment the patch offers, the smaller it is. Each occupied patch $j \neq i$ contributes to the colonization probability on patch i in an additive manner, so that the invasiveness S_i of the whole metapopulation with regard to patch i is $S_i = \sum_j o_j S_{ij}$, where o_j is the occupancy of patch j: it is one if j is occupied and zero otherwise. The colonization probability C_i of an empty patch i is assumed to be $C_i = S_i^2/(S_i^2 + y^2)$, where y is a constant that can be estimated from patch occupancy data by non-linear regression. Assuming that the probability E_i of local extinction is a negative power function of patch size: $E_i = eA_i^{-x}$ (x is also a fitted constant), the model can be implemented either as a stochastic simulation for a finite set of habitat patches, or as a Markov chain which incorporates the **pairwise biotic isolation structure** of the metapopulation. The approach can be extended to two-species competitive interactions with the same assumptions as in Caswell and Cohen (1991b): one species dominates the other, which is a better disperser in exchange. The model was applied to a competitive metacommunity of two butterfly species which differed only in competitive strength and dispersivity – all other constants of the model were non-linear regression parameters obtained from a real patch-occupancy data base of a butterfly species (Hanski et al., 1995). The question was that originally posed by Nee and May (1992), who used a similar, but spatially implicit model: does permanent habitat destruction (removal of a fraction $1 - h$ of the patches) help the fugitive species to persist? As in Nee and May (1992), the answer was yes within a wide range of h for almost any patch configuration. Yet, the coexistence of the two species was possible only in a small interval of h, outside of which either or both species died out from the habitat patchwork (see also Tilman et al., 1994).

The basic assumptions of island biogeographic and metapopulation modelling are similar in that the relevant processes of both are local extinctions and colonizations by immigration. Equations (4.1) and (4.21) represent two endpoints of a continuum of possible metapopulation structures, one with a set of habitat patches of similar size and quality, which are connected in every pairwise combination through interpatch dispersal, the other with a practically infinite external source of migrants, but no significant interpatch dispersal connections among islands of variable size and habitat quality. Clearly most real situations fall in between these extremes (Hanski, 1991a), showing a variable patch size or patch quality distribution, and a certain pattern of partial interpatch connectance and mainland colonization effects.

There are, on the other hand, obvious differences between the metapopulation approach and that of classical island biogeography, stemming from the difference in the questions that these disciplines ask. Island biogeography models were originally designed to explain the commonly observed log–log linearity of the species–area relationship, a large-scale pattern in nature that seems to be surprisingly robust. The variable that island biogeography is ultimately interested in is therefore the **number** of co-occurring species as a function of island area – the actual species compositions of the assemblages are disregarded. This means omission of even the combinatorial information of local species compositions, which is essential to consider in metacommunity models. The classical multispecies model of MacArthur and Wilson (1967) in fact assumes the coincidence of many dynamically independent – that is, not interacting – species, each having its own incidence function dependent on island area and distance from the mainland. The local communities, and also the metacommunity thus constructed, are of a random composition in the species–area model. Many aspects of the species–area relationship could be explained with these assumptions on the biogeographic scale, but species interactions are indispensable to consider for smaller-scale problems of coexistence to understand. Metapopulation models seem to be quite well suited for that purpose.

4.3 AGGREGATION MODELS OF SPECIES INTERACTIONS

Besides reducing the representation of local dynamics, the crucial simplifying assumption of the metapopulation approach as compared to patch abun-

Aggregation models of species interactions

dance models is the omission of the actual spatial allocation of patches, thus making the spatial reference implicit in the model. This modelling strategy works for single species and for a small number of interacting species as well. Under certain constraints on the structure of the model, we can go even further with the simplification of our assumptions regarding the spatial structure of the habitat. Specifically, as far as only pairwise interactions of species are concerned, one can even give up the assumption of a necessarily patchy habitat structure and consider inhomogeneity in the spatial distribution of the species in a rather general sense, without any direct reference to space. The trick is to assume that **biotic interactions** are inhomogeneously distributed, so that the probability of interaction between individuals of the two species might be locally disproportional to the product of the actual overall abundances.

The simplest such system considers spatially aggregated interactions of parasitoids with their hosts; it is a modification of the classical Nicholson–Bailey (1935) model. The basic idea of including non-random interactions into the Nicholson–Bailey system was proposed a long time ago by Bailey *et al.* (1962), and Griffiths and Holling (1969). The idea developed into a ramified family of models, assuming some kind of patchy habitat structure first in a 'semi-explicit' spatial setting, by the specification of the joint distributions of the interacting populations on the patchwork habitat (e.g., Hassell and May, 1973, 1974). The literature of aggregated host–parasitoid interactions has grown very large since then, leading ultimately to the spatially implicit phenomenological model of May (1978), which accounts for non-random parasitization regardless of the spatial mechanisms that actually realize it. Shorrocks *et al.* (1979), Atkinson and Shorrocks (1981) and Ives (1988, 1991) adapted the same idea to competitive situations. Pacala *et al.* (1990) review some aspects of host–parasitoid aggregation, and the early monograph by Hassell (1978) and a recent one by Godfray (1994) present a broad spectrum of both spatial and non-spatial extensions to the classical Nicholson–Bailey model, with field examples and further references.

A less well-known class of single-species models, assuming spatial population structure in an implicit manner, is that of spatial inhibition processes (Matérn, 1960; Strauss, 1975; Ripley, 1977; Ripley and Kelly, 1977; see Kenkel, 1995, for a review of actual and possible applications to clonal plant growth). Spatial inhibition models are a specific type of stochastic point processes with a small range of biological applicability (see Kenkel, 1993, for examples); I do not treat them in more detail here.

4.3.1 The non-spatial reference: the Nicholson–Bailey model

Most parasitoid species are ichneumonid insects, which lay eggs into the individuals of other insect species, during the larval or pupal stage of the host's development. Parasitized host instars then give birth to imagos of new parasitoids, instead of their own kind. In a sense, parasitoidism is a delayed version of instar predation; it leads to the ultimate death of the host larva before it could develop to become an imago and to reproduce. The main difference from predation is that a host instar can be the target of more than one parasitoid attack consecutively, but it can yield only a limited number of parasitoid imagos in the end.

The original Nicholson–Bailey model regards this situation, when both the parasitoid and the host are annual species, and parasitoid attacks are randomly distributed on the host individuals. The model was formulated as a very simple discrete-time system, one time unit representing one year:

$$H_{t+1} = FH_t \exp(-aP_t)$$
$$P_{t+1} = cH_t[1 - \exp(-aP_t)],$$
(4.23)

H_t denotes the number of host larvae susceptible to parasitization, P_t is the number of parasitoid imagos within the habitat at time t. F is the fecundity of the host – that is, the number of female offsprings a single host female can produce in the absence of the parasitoid – and c is the number of parasitoids emerging from a parasitized host instar, irrespective of how many times it was attacked. F can be expressed in exponential form as $F = \exp(r)$, where r is the intrinsic rate of increase for the host. The parameters of (4.23) can be rescaled so that $c = 1$, which I assume to hold in the forthcoming. The exponential term in both equations represents the probability that a host is not parasitized at all. This particular form comes from the assumption that the hosts are encountered at random: aP_t is the expected number of attacks on a host if a is the fraction of the total area of the habitat that a parasitoid female can search for hosts in a lifetime, and parasitoids search independently from each other. a is the efficiency of the parasitoid in exploring susceptible hosts. In mathematical terms, $\lambda = aP_t$ is the parameter of a Poisson distribution on the number of attacks per host, so that the probability of zero attacks (that is, escape from parasitization) for a host larva is the zero term of the Poisson series:

$$p_0 = \frac{\lambda^0}{0!} \exp(-\lambda) = \exp(-aP_t).$$
(4.24)

The probability of getting parasitized at least once is $1 - p_0$.

Equation (4.23) admits a single unstable interior fixed point with divergent oscillations. This is not a surprise given that the system is in fact a discrete time model for a specific type of predator–prey interaction, with assumptions similar to those of the time-continuous Lotka–Volterra model (3.22). Note that both these models lack density-dependent self-regulation, the only limiting factor for each species being the abundance of the other. The corresponding Lotka–Volterra model is neutrally stable, a little better behaved than (4.23). The instability of the Nicholson–Bailey model is attributable to the inherent time lag in population responses due to the discrete time scale.

The similarity of the two models go even further: for example, the effect of host (or prey) density dependence has a stabilizing effect on both. The Lotka–Volterra model (3.22) becomes asymptotically stable with damped oscillations, if a finite carrying capacity K for the prey is introduced. A density-dependent version of the Nicholson–Bailey model is obtains if the exponential fecundity term $F = \exp(r)$ of the host is replaced by a logistic term $F(H_t) = \exp[r(1 - H_t/K)]$ to yield

$$H_{t+1} = H_t \exp\left(r\left(1 - \frac{H_t}{K}\right) - aP_t\right)$$
$$P_{t+1} = H_t[1 - \exp(-aP_t)].$$
(4.25)

Beddington et al. (1975) studied this model, showing that it is more stable than the original (4.23) system in the sense that its trajectories are all bounded. Within the bounds, however, one might expect an almost unimaginably complex dynamics from (4.25) with good reason, knowing that even very simple single-species, density-dependent, discrete time models like (3.31), (3.34) or (3.35) can produce almost anything, ranging from fixed point stability to chaos. The same applies to this system, which actually returns (3.34) exactly if $a = 0$. However – at least within the biologically reasonable part of the parameter space – the overall effect of assuming host density dependence on the dynamics of (4.23) is stabilizing in the above sense, as shown on Figure 4.9.

In Figure 4.9, one can easily compare the dynamics of (4.25) to that of its single-species equivalent (3.34). Note that the only interspecific interaction parameter of (4.25) is a, the searching efficiency of the parasitoid, which is responsible for the depression of the equilibrium host abundance, H^*, below K, the equilibrium abundance without parasitization (that is, when $a = 0$). Figure 4.9 thus demonstrates that the efficiency of the parasitoid in decreasing the equilibrium host abundance is of decisive importance in determining the stability of the system. The effect of parasitoid action on stability is not monotonic: moderate parasitoid control stabilizes, but too strong an interaction destabilizes the host–parasitoid coexistence.

The numerical investigation of a part of the parameter space shows that the system produces a similarly complicated scenario of bifurcations (ending in the chaotic region) along the host reproductive rate axis as does the single-species model (3.31), but the chaotic behaviour emerges at different fecundities, depending on the actual value of a. Compared to (3.31), the critical value of the reproductive rate at which the equilibrium point loses stability is larger for values of the interaction parameter a that do not depress the equilibrium abundance H^* of the host below half of its carrying capacity, K. Larger a values (where H^*/K is smaller than 0.5) are

144 Spatially implicit patch models

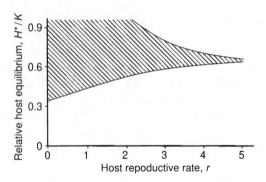

Figure 4.9 Stability map of (4.23). The fixed point is stable inside the hatched area only. Note that parasitization levels (that is, a values) depressing H^* below $K/3$ are always destabilizing, but less efficient parasitoids stabilize the dynamics of the host population at moderate rates of host reproduction.

destabilizing and below $H^*/K \approx 0.36$, there is no r at which the fixed point is stable. The presence of a moderately efficient parasitoid stabilizes the dynamics of the host population.

Hassell and May (1973) review a number of other non-spatial models, each one a modification of the Nicholson–Bailey system, incorporating different forms of density dependence affecting the interaction term, that is, the probability of encounters instead of the fecundity of the host. These models are in fact out of the scope of this book, being essentially non-spatial, but their results also demonstrate the structural similarity of the Lotka–Volterra system with the Nicholson–Bailey model, which I will also refer to in a spatial context later. This is the reason for mentioning them here briefly, without going into details. One possibility is the inclusion of the functional response of the parasitoid to host density in the form of the disque equation (3.25) into the exponential term, which proves to be destabilizing just as in the Lotka–Volterra model. The other assumes a decrease in the searching efficiency of the parasitoid with increasing density, that is, an indirect postulation of parasitoid interference or intraspecific competition. This is stabilizing for the system in a large part of its parameter space, just as is predator density dependence in the Lotka–Volterra model.

4.3.2 Spatial heterogeneity included: aggregation of encounters in a patchy host distribution

The Nicholson–Bailey model, and also its density-dependent version (4.25), assume random encounters of hosts and parasitoids, which amounts to postulating the spatial independence, and thus the overall spatial homogeneity, of interactions. The implicit inclusion of heterogeneity (spatial or otherwise) is possible through appropriate modifications to the specific random encounter assumption of the non-spatial model. One simple example is assuming that a fixed proportion α of the hosts are always safe from parasitism, and all parasitoid attacks are distributed at random on the rest. This refuge effect is easily built into the Nicholson–Bailey model, to yield

$$H_{t+1} = \alpha F H_t + (1-\alpha) F H_t \exp(-aP_t)$$
$$P_{t+1} = (1-\alpha) H_t [1 - \exp(-aP_t)].$$
(4.26)

Note that here it is assumed that the environment consists of safe sites and sites exposed to parasitoid search, and that the whole host population is redistributed among these sites in proportions α, $1 - \alpha$ in each time unit. Hassell and May (1973) studied this model as a limiting case of a more general system ((4.27) below), and found that its stability properties are somewhat better than those of the Nicholson–Bailey model (Figure 4.10). There is a narrow region of the parameter space where the unique interior equilibrium is stable. Below the stable region there is no equilibrium at all, however, because too large a proportion of the hosts are always out of the control of the parasitoid; thus both populations increase infinitely. Above the stable region, the feasible fixed point exists, but it is unstable, just like in the original model: there are not enough safe sites to change the dynamics qualitatively.

The more general version of (4.26) assumes n patches, on which both the host and the parasitoid populations are distributed, so that on patch i a proportion α_i of the full host larva population, and a proportion β_i of the parasitoids, are present. Inherent in this is the assumption that the newborn of both populations are redistributed on the patches in these proportions every generation (Murdoch and Oaten, 1975). It is not easy to interpret this assumption in biological terms, since it requires either the ability from both

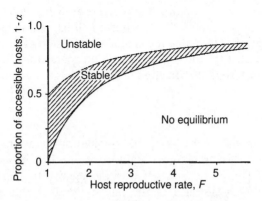

Figure 4.10 Stability map of the refuge model (4.26). The shaded area is the stable domain of the parameter space. (After Hassell and May, 1973.)

146   Spatially implicit patch models

species to measure all local densities simultaneously somehow and chose habitat accordingly, or a mechanism that ensures redistribution automatically. Both are possible in principle, but field examples might be hard to find for any of them. Therefore it seems better to think of this assumption as one of the many possible representations for the aggregation of parasitoid attacks on the hosts. Encounters are random within the patches, according to the Poisson distribution as before. With these assumptions, the general model looks like

$$H_{t+1} = FH_t \sum_{i=1}^{n} \alpha_i \exp(-a\beta_i P_t)$$
$$P_{t+1} = H_t \sum_{i=1}^{n} \alpha_i \left[1 - \exp(-a\beta_i P_t)\right].$$
(4.27)

Equation (4.26) is a specific form of this system, formulated for the sake of mathematical convenience with two patches only, one for the host refugees, the other as an arena for interaction. In that particular case,

$$\alpha_1 = \alpha, \quad \alpha_2 = 1 - \alpha, \quad \beta_1 = 0, \text{ and } \beta_2 = 1.$$

In the general case with many patches, the spatial distributions of both the host (α_i) and the parasitoid (β_i) must be specified somehow. This can be done by using empirical (or arbitrary) distributions for both species, which means introducing a lot of parameters into the model, and, as a direct consequence, losing much of its theoretical value. The stability condition for (4.27) is known in analytical form (Hassell and May, 1973), but it is not very revealing as it stands:

$$F \sum_{i=1}^{n} \left[\alpha_i \left(a\beta_i P^* \right) \exp\left(-a\beta_i P^*\right) \right] < \frac{F-1}{F};$$
(4.28)

P^* is the parasitoid abundance at the unique interior equilibrium of (4.27). The only obvious conclusion one might arrive at by inspecting (4.28) is that the stability of the system (4.27) depends on all its parameters.

The other option is to define the spatial distribution of the host arbitrarily, but to suppose that the behaviour of the parasitoid depends on the pattern of the host, so that parasitoid distribution is a function of α_i. Hassell and May (1973) used

$$\beta_i = k\alpha_i^\mu$$
(4.29)

to define the response of the parasitoid population to the spatial pattern of the host, in which k is a scaling constant to ensure that the values of β_i sum up to one; $k = [\Sigma \alpha_i^\mu]^{-1}$. μ is the 'parasitoid aggregation index', which is interpreted as the preference of the parasitoids to more abundant host patches. If $\mu = 0$, there is no density response at all; parasitoids distribute evenly and independently from the host pattern. Then the model is essentially identical to the Nicholson–Bailey form assuming random search, and, of course, it gives the same result (an unstable fixed point). If $\mu = 1$, $\beta_i = \alpha_i$ for all i, that is, the spatial distribution of the parasitoid population follows exactly that of the host. With $\mu > 1$, the parasitoid has a disproportional preference to dense host patches; $\mu \to \infty$ represents extreme propensity for aggregation in good patches, up to the point where the only host patch that parasitoids visit is the most abundant one; all the others are refuges for the host. Figure 4.11 shows the density response of the parasitoid for some values of the aggregation index.

Now it is possible for any specific host distribution α_i to calculate the stability boundaries of the stability map on the $F - \mu$ parameter plane of (4.27). Hassell and May (1973) gave an example, using a particular type of heterogeneous host distribution with one patch of high, and $n - 1$ patches of low host density, where a fraction γ of the total host population was allocated to the single high-density patch, and the low-density patches shared the remaining in equal proportions, $(1 - \gamma)/(n - 1)$ each. Note that both $\gamma = 0$ and $\gamma = 1$, and also $\gamma = 1/n$ correspond to the complete spatial homogeneity of the populated patches. According to (4.29), parasitoid aggregation cannot occur at these values of γ, where (4.27) again becomes non-spatial, therefore unstable, like the Nicholson–Bailey model. Note that $\gamma > 1/n$ is necessary for the single patch to be called 'high-density' by reason.

The stability maps of this system for some combinations of μ, F, n and γ are shown on Figure 4.12, from which the main conclusions are quite straightforward to draw: parasitoid aggregation on the high-density host patch (increasing μ) has a stabilizing effect on the interaction, but this

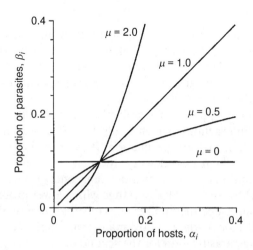

Figure 4.11 Relationships between the proportion of hosts and that of searching parasites in the ith patch, for different values of the parasite aggregation index μ, calculated from (4.29). (After Hassell and May, 1973.)

148   Spatially implicit patch models

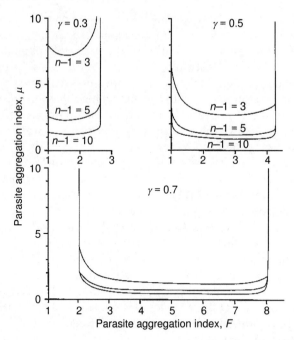

Figure 4.12 Stability boundaries for the model ((4.27), (4.29)) within the parameter space spanned by parasite aggregation index (μ), host rate of increase (F), number of 'low host density' patches ($n-1$), and proportion of hosts in the 'high host density' patch (γ). F and μ values above the u-shaped boundary lines allow for stability. (After Hassell and May, 1973.)

depends on the aggregation of the host, itself in turn clearly related to γ and n. Intermediate values of γ (that is, those in between $\gamma = 1/n$ and $\gamma = 1$, but sufficiently far from both) produce wide regions of stability on the $F - \mu$ map, which also expand downwards with increasing $n - 1$, the number of scarcely populated patches.

Thus the ultimate reason of stability seems to be the aggregation of host–parasitoid interactions, which leads to differences in the levels of parasitization within the host population. If the parasitoid tends to aggregate on patches of high host density, the hosts on these patches will be heavily parasitized, many of them more than once, whereas hosts on the low-density patches are in a relative refuge with a lower probability of being encountered at all. Since the number of parasitoids emerging from a single host individual is independent of how many times (above one) the host was encountered, any parasitization for the second or more time is in vain – the eggs are misplaced from the point of view of parasitoid population growth. In other words, the parasitoids are in fact short of hosts in the high-density patch, leading to a high level of apparent parasitoid autocompetition on that locality (Free et al. (1977) call this phenomenon the 'pseudo-interference' of the parasitoids). If we now compare this result to that of the continuous time predator–prey Lotka–Volterra model with predator autocompetition, the correspondence is again obvious: intraspecific competition stabilizes the interaction in both model types.

4.3.3 Spatially undetermined aggregation of interactions

Accepting that it is, in fact, the distribution of parasitoid attacks on the hosts that matters regarding stability, one might come to the conclusion that it is not even necessary to specify the spatial structure of the habitat to construct models of aggregated interactions. It is sufficient to assume that parasitoid encounters are not Poisson distributed, but follow some other distribution which can account for the clumping of parasitization events on the hosts somehow. Of course the mechanism leading to encounter aggregation remains cryptic in such a model, leaving the possibility of different interpretations open. Assuming a negative binomial distribution of parasitoid attacks on host individuals instead of the Poisson distribution that implies completely random encounters, May (1978) modified the Nicholson–Bailey model, following the line initiated by Griffiths and Holling (1969). He replaced the zero term of the Poisson series with that of a negative binomial of expected value aP_t and parameter k, so that the probability of escape from parasitization became

$$p_0 = \left(1 - \frac{aP_t}{k}\right)^{-k}. \tag{4.30}$$

k is the 'clumping parameter' of the negative binomial. In the present context, k is a measure of the aggregation of parasitoid attacks on the host individuals. Its role in the distribution is clear from the relation

$$CV^2 = \frac{\mathrm{var}(l)}{\mathrm{mean}^2(l)} = \frac{\mathrm{var}(l)}{(aP_t)^2} = \frac{1}{aP_t} + \frac{1}{k}, \tag{4.31}$$

where CV is the coefficient of variation for the distribution of l, the number of attacks on the hosts. Large values of CV correspond to clumped distributions, so that the smaller k is, the more aggregated the parasitization events on the hosts. For finite positive values of k, more hosts will be attacked more often and more will be not encountered at all than in the Poisson case, which yields from (4.30) in the limit as $k \to \infty$. The model now takes the form

$$H_{t+1} = FH_t(1 - aP_t/k)^{-k}$$
$$P_{t+1} = cH_t\left[1 - (1 - aP_t/k)^{-k}\right] \tag{4.32}$$

150   Spatially implicit patch models

This system admits a single interior equilibrium point, which is locally stable if and only if $k < 1$. The dependence of the equilibrium parasitoid pressure aP^* on k, and the stability of the fixed point of (4.32), is shown in Figure 4.13. Note that $k = 1$ corresponds to attacks being geometrically distributed. This borderline case already represents a certain degree of clumping, so that the condition for stability is not simply aggregation, but **strong enough** aggregation of parasitoid attacks, itself effective through the consequent increase in parasitoid pseudo-interference as explained above.

One possible (in this case, spatial) model situation leading to the negative binomial distribution of attacks on host individuals is when the habitat consists of many patches, with inhomogeneously distributed prey and a spatially correlated, gamma-distributed parasitoid population, if their local interactions within the patches are exactly of the Nicholson–Bailey type. May (1978) also notes that different assumptions regarding the actual spatial patterns of the two populations will lead to different distributions of encounters, but as far as only the clumping of parasitoid attacks is concerned, the negative binomial can always be a good phenomenological approximation.

Chesson and Murdoch (1986) show that assuming random spatial aggregation of the parasitoid, independent from the density pattern of the host, leads to the same model (4.32). This again supports the conjecture already exposed that it is the pseudo-interference of the parasitoids due to encounter aggregation that has the stabilizing effect, whatever actual mechanism it

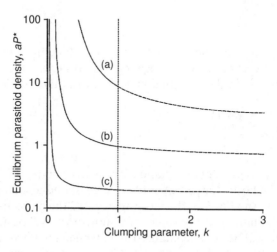

Figure 4.13 Equilibrium parasitoid pressure (aP^*) and the stability of the equilibrium of (4.32), as a function of the 'clumping parameter' (k) and host fecundity (F). (a) $F = 1.2$; (b) $F = 2.0$; (c) $F = 10.0$. The solid part of the curves correspond to stable equilibria, the dashed parts correspond to unstable equilibria with divergent oscillations (after May (1978)).

is produced by. In fact, even negative spatial aggregation, that is, aggregation of parasitoids on low-density host patches, will obviously cause parasitoid pseudo-interference, and thus it can promote stability (Hassell and May, 1988; Chesson and Murdoch, 1986). Based on these, and a number of other mechanisms, a rather general and attractively simple approximate stability criterion was deduced by Hassell and Pacala (1990) and Hassell *et al.* (1991). They state that if the coefficient of variation for the distribution of parasitoid attacks on the hosts exceeds unity, that is, if $CV > 1$, the interaction can be expected to be stable. Whether the actual mechanism realizing this state is positively or negatively density-dependent or density-independent does not matter from the viewpoint of stability, since the effect is bound to the pseudo-interference of the parasitoids, which is a direct result of aggregated parasitoid attacks.

It is interesting to compare these results to those of the patch-abundance model of predation by Murdoch *et al.* (1992) (cf. Section 3.4.2), who conclude that preferential migration by both prey and predator might be even destabilizing in a two-patch system, provided – realistically – that the prey population prefers the less crowded prey patch, whereas the predator seeks the one with a high abundance of prey. Obviously both these preferences tend to eliminate the difference in prey abundance, that is, prey aggregation, on the two patches. If prey abundance is levelled off on the patches, no prey-density-dependent predator aggregation is possible, so the originally aggregated interaction of the two populations becomes homogeneous in space, and thus unstable in time. This cannot happen in the host–parasitoid models above, since it is assumed throughout that the aggregated distribution of prey is fixed forever by some external mechanism; aggregation is not dissolved by the autonomous dynamics of the system.

4.4 SUMMARY

Metapopulation and aggregated interaction modelling represent two different ways of attempting to solve the notorious methodological problem of mathematical intractability in patch-abundance models with a large number of habitat patches. Metapopulation (or patch-occupancy) systems solve the problem by simplifying the representation of within-patch dynamics considerably, compared to patch-abundance models. The habitat is assumed to consist of a very large number of patches, each of which undergoes repeated cycles of local extinction and recolonization by migrant propagules. Within-patch dynamics is implicitly represented through the parameters of local extinctions and colonizations; making them dependent on the actual number (or fraction) of occupied habitat patches enables habitat size and quality effects and the rescue effect to incorporate into metapopulation models. Such simple modifications proved successful in explaining certain patterns of abundance on the field, an example of which is the core-satellite distribution of species on patchy habitats. The predictions of patch-abundance models and metapopulation systems tend to be similar: habitat heterogeneity is beneficial from the viewpoint of population dynamical stability. These predictions originate from different mechanisms in the two, however: patch-abundance models are equilibrium systems basically, whereas metapopulation models are inherently non-equilibrium, at least on the local (patch-level) scale. The stabilizing effect can be attributed mainly to the spatiotemporal heterogeneity assumed in both: if the spatial distribution of the habitat is patchy, and local dynamics are asynchronous on the patches, then the system can be expected to be stable on the regional scale.

Discrete time host–parasitoid models are a more specific type of spatially implicit systems; in fact these are variants of the classical Nicholson–Bailey model, in which the aggregation of parasitization events is assumed. The effect of aggregation is embedded in the interaction parameters, without any explicit reference to the cause or the spatial mechanism of aggregation. The aggregation of interactions leads to the pseudo-interference of parasitoids, and thereby to an increased stability of the system, irrespective of the kind of density-dependent or density-independent mechanism maintaining the aggregation of interactions.

Part Two
Neighbourhood Models of Population Interactions

5 Site-based neighbourhood models

5.1 INTRODUCTION

During about the past ten years, one of the most widely used and most successful methodologies of spatiotemporal modelling has certainly been that of cellular automata (CA) in quite a few scientific disciplines. The method is useful and popular at the moment, but the widespread application of CA in modelling the dynamics of complex physical, chemical or biological objects is a relatively new achievement, compared to the long tradition of spatial mass-interaction approaches like reaction–diffusion systems or patch models. Although the very first CA appeared in the 1940s as conceptual models for deterministic self-replicating systems (von Neumann, 1966), actual realizations were practically impossible at that time due to the lack of substantial computing capacity. It is just an interesting coincidence that the theory of deterministic CA made a significant contribution to the solution of the computing capacity problem: the second major field of its application was modelling devices capable of universal computation – in fact these were, in a sense, theoretical pre-images of digital computers (Wolfram, 1986; Toffoli and Margolus, 1987). The theory of automatic computations did not induce practical applications for quite a few years, however; microelectronic technology had to develop for more than a decade to reach a level that enabled the idea to 'materialize'. Now that fast personal computers are generally available, experimentation with huge CA became practicable and the way for scientific applications is open.

Besides modelling spatiotemporal dynamics in actual physical, physicochemical and biological systems (Wolfram, 1986; Hogeweg, 1988; Comins *et al.*, 1992), CA have also had a spectacular career as fashionable solitary games in popular scientific journals ever since the early seventies (Gardner, 1970), when Conway's 'Game of Life' (cf. Section 5.2.1), one of the simplest and definitely the most popular such CA, appeared.

In the ecology literature, discrete-space finite-state neighbourhood models are called CA in general. In the strict sense, however, CA are deterministic in their rules of local interaction (Wolfram, 1986), whereas most ecological applications are probabilistic, belonging in fact to a broader type of finite-state models called **interacting particle systems** (IPSs) (Durrett, 1988; Durrett and Levin, 1994a, b), which includes CA as special cases. IPSs are more general than CA in two fundamental respects: they allow for (i)

156   Site-based neighbourhood models

stochastic and/or (ii) asynchronous interaction rules. Both these differences will be explained in Section 5.2, along with a general characterization of IPSs; Section 5.3 includes five different population dynamical applications.

5.2   INTERACTING PARTICLE SYSTEMS AND CELLULAR AUTOMATA

The name 'interacting particle system' comes from the pioneering applications of stochastic discrete-event models in physics (e.g., spin-glass models; Wolfram (1986)) to systems in which the interacting units were spatially fixed particles. The particle-to-site correspondence is perfect in such physical models: it was unnecessary to emphasize that the interacting objects of CA and IPSs are in fact **spatially** defined. In ecological applications, the individual-to-site correspondence is fuzzier; it can change with time, for example, and the spatial scale of an individual and a site need not be the same. The ultimate object of interaction is obviously the site in ecological IPSs, therefore the name of the model class is somewhat misleading in ecological context.

5.2.1   The structure of interacting particle system models

The complete specification of an IPS model requires the definition of

1. a regular lattice of sites (cells);
2. a finite – typically small – set of possible states of the cells;
3. the size and shape of the neighbourhood, that is, the set of cells affecting the next state of the focal cell;
4. a next-state function (also called 'state transition rule' or 'updating rule'); and
5. an initial state configuration.

Point 1

The lattice might be of any (integer) dimensionality d in principle, but for spatial population models, $1 \leq d \leq 3$ is a straightforward limitation. Any topologically uniform subdivision of the space with the given dimensionality defines a proper lattice; in two dimensions – which is the most frequent choice in existing ecological applications – it may be a square or a hexagonal grid (Figure 5.1), usually of infinite extension in theory, but always a finite one in numerical practice. The opposite edges of finite grids are sometimes merged, so that the lattice forms a torus in a topological sense. The imposition of such a 'cyclic boundary condition' or 'wrap-around margins' is a useful practice to avoid edge effects, which are inevitable otherwise, but it can introduce artefactual spatial periodicity into the system in exchange. Increasing the size of the grid usually helps to weaken both these effects, but

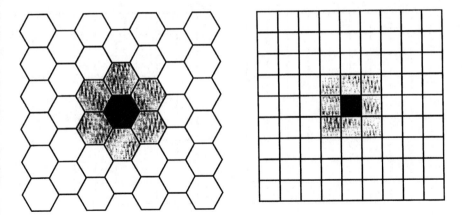

Figure 5.1 Square and hexagonal lattice. Grey sites are potential nearest neighbours of the black site.

it requires a dramatic increase in computing capacity (according to a power function with the exponent close to d, the dimensionality of the lattice), in terms of both data storage and CPU time.

Point 2

Each site i is in one particular state s_t^i from a finite set of i-states at every discrete time t. Apart from spatial position, which is obviously different for different sites, the relevant i-states are usually specified by a very small number of i-state descriptors, each taking only a few values. The simplest possible i-state consists of one Boolean i-state descriptor, whereby any single site can be in one of two states, 1 ('on') or 0 ('off') at any time, representing the presence or the absence of an individual within the cell, for example.

Point 3

The neighbourhood definition specifies for each site which other sites determine its state in the next time unit. Neighbourhood topology is space- and time-invariant in most ecological applications of CA and IPSs, the most frequently used definitions being one of the two basic types of 2D square lattices illustrated in Figure 5.2. Space invariance means that the size and the shape of the neighbourhood (relative to the focal site) does not change from site to site, and time invariance means that it is constant in time. Both these simplifying assumptions can be dropped in actual models, if necessary, to represent the spatial or temporal inhomogeneity of the environment, for

158   Site-based neighbourhood models

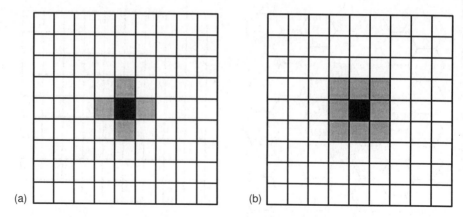

Figure 5.2 The most frequently used IPS neighbourhoods in square lattices: (a) Neumann neighbourhood; (b) Moore neighbourhood.

example. In population dynamical models, it is also possible that the size and the shape of the neighbourhood depends on the *i*-state of the focal site: this can be the case, for example, if the size of the effective neighbourhood around the home site of an animal depends on its age – older individuals can control a larger neighbourhood.

Point 4

The state s_{t+1}^i of cell *i* at time $t+1$ depends on the state configuration Θ_t^i of the cells within its neighbourhood (including the focal cell *i* itself), at time *t*. The next-state function $f(\Theta_t^i)$ specifies the rules of the state transition for every possible neighbourhood configuration, either in the form of an explicit list coupling neighbourhood configurations to focal cell next states (that is, a list of all possible $\Theta_t^i \to s_{t+1}^i$ pairs), or as an algebraic or logical expression. An explicit list is practically inconvenient to use, if the neighbourhood is not very small (that is, if it consists of more than, say, five cells). Conway's 'Game of Life', for example, is a 2D Boolean CA with the Moore (that is, the nine-cell) neighbourhood, for which the next-state function can be easily formulated in terms of the number of surrounding 'on' cells: the focal cell remains in its previous state if it had two 'on' neighbours; it turns 'on' anyway if it had exactly three, and it turns 'off' otherwise. These simple, verbally formulated conditions are equivalent to a highly uninstructive list of $2^9 = 512$ elementary rules of the $\Theta_t^i \to s_{t+1}^i$ type, which is actually one of the possible $2^{512} \approx 1.34 \times 10^{154}$ such lists (rules) applicable to 2D Boolean CA with Moore neighbourhoods.

Deterministic CA, like the 'Game of Life', apply deterministic transition rules only: a certain neighbourhood configuration yields a certain next state of the focal cell with a probability one. Probabilistic CA differ from these in

Interacting particle systems and cellular automata

admitting stochastic updating rules, which specify the probability distribution of next states for each possible neighbourhood configuration. In other words the same neighbourhood pattern can lead to different next states of the focal cell, each of the possible outcomes being realized with a specified probability.

The updating rules of CA (either deterministic or stochastic) are synchronous in the sense that the whole lattice is supposed to flip to the next-state pattern instantaneously, the next state of each cell depending on the previous configuration of its own neighbourhood. It is possible to apply another procedure, namely updating the cells one by one, in a predetermined or in a random order, in which case the IPS we obtain is not even in the loose sense a CA any more. Random sequence updating introduces a third possible source of stochasticity into the dynamics of the system, besides probabilistic updating rules and random initial configurations. This modification might change the dynamics of certain models considerably, and in the limit it leads to continuous time IPS models (Durrett and Levin, 1994b), usually a little more accessible for analytical treatment than synchronously updated CA. Most ecological applications of IPSs that have appeared up to now are synchronously updated stochastic CA, with a very limited number of exceptions (cf. Durrett and Levin, 1994b).

Point 5

The initial state configuration can be a fixed, regular or irregular pattern, or a random one. In any case, the temporal dynamics of the pattern are determined if the state transition rules are deterministic. The trajectories of many IPSs, and especially of many CA, are highly dependent on initial patterns, that is, on either or both the densities and the actual configurations of initial site states.

5.2.2 Mean-field and configuration-field approximations to interacting particle systems

As might be seen from the foregoing description, the formal structure of IPS models is very simple on the local (site) level. The constraints and the rules of their local behaviour are easy to state, easy to understand, easy to implement on a computer and easy to change, but the consequences of the local structure on the formation of lattice level patterns is almost always impossible to infer by analytical means. This is true in the literal sense regarding deterministic CA, for which analytical approaches are nearly out of the question. The updating rules and the initial configuration of a CA tell almost nothing about how the system will behave if left alone for a number of generations, except if it follows very specific or rather uninteresting rules like unconditional extinction (whatever the neighbourhood configuration was, the focal site switches 'off'). One exception, a specific updating rule for

160 Site-based neighbourhood models

Box 5.1 Self-replication of patterns with the parity rule in CA (Fredkin (1970) described by Toffoli and Margolus (1987))

Consider the 2D Boolean CA on a square lattice with a Neumann neighbourhood. The updating rule is very simple: if the number of 'on' cells in the neighbourhood (including the focal cell) was odd, the focal cell turns 'on'; if it was even, the focal cell turns 'off'. This so-called **parity rule** produces identical copies of any bounded pattern in a finite number of iterations; the daughter copies inherit the orientation of the parent as well. Then the copies replicate further, launching a temporally supralinear process of 'population growth' and a spatial spread of

$t = 0$

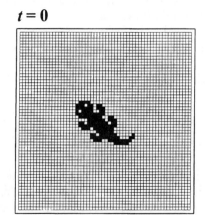

$t = 15$

$t = 16$

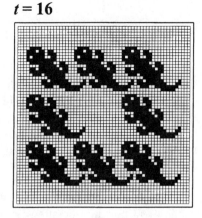

Figure [5.1] 'Self-replication' with the parity rule in a 2D Boolean CA with Moore neighbourhood.

> the copies within the infinite lattice. The reason why the growth of the population is not exponential is the mutual quenching of 'individuals' due to spatial interference, which occurs if an even number of them happen to occupy the same position. With a Neumann neighbourhood, the number of daughter copies in an elementary cycle of self-replication is four; with a Moore neighbourhood, it is eight (Figure [5.1]). The duration of a replication cycle depends on the size of the parent pattern.
>
> This is an extremely simple CA system, one of the rare exceptions in which an emergent property can be predicted from the algebraic formulation of the rule. The self-replication property can be proven rigorously for the parity rule, because its algebraic form is linear. The self-reproducing property of this system is extremely sensitive to any change in the updating rule: assuming a slight stochasticity or asynchrony in the updating algorithm destroys self-replication completely and very quickly. This is a strong reason to conclude that such a self-reproducing CA can be at most a metaphoric representation of a property that is characteristic to life in a much more complicated and robust manner (Boerlijst, 1994).

which it is possible to infer the outcome without numerical simulation, but which nevertheless is interesting, is the so-called **parity rule**, which is capable of multiplying any initial pattern (Box 5.1).

There are rough characterizations for the possible outcomes of some elementary types of deterministic CA (Wolfram, 1986), based mostly on patterns produced by the simplest possible system (one dimensional, deterministic, Boolean, two-neighbour), but there is no obvious correspondence between the type of the updating rule or that of the initial configuration and the type of the actual outcome. One has to let the automaton work to see what state configuration it evolves to from a given initial pattern. The global (lattice-level) properties of CA are not transparent from, albeit they are of course determined by, the local rules.

Given the difficulties of almost any analytical approach to deterministic CA, surprising though it may be, nevertheless it is true that there exists an increasing arsenal of sensible analytical methods for the study of stochastic IPSs. It is also true in general, however, that most theorems on IPSs concern oversimplified cases from the viewpoint of ecological applications (but see Durrett and Levin (1994b)) and references therein, for exceptions where the existing body of theory could, in fact, have been applied to models that were studied numerically by their authors originally). One of the most effective simplifications, frequently leading to some analytical tractability at least, is what physicists call the **mean-field approximation**. This means ignoring the actual neighbourhood configurations in the next-state function, and assuming that the next state of each cell depends on the expected (average) neighbourhood (mean-field) state only. This simplification eliminates the spatial information from the system altogether, and it degenerates the model from an object-interaction (i-state configuration) to a mass-interaction (i-state distribution) type: it becomes either non-spatial or spatially implicit. Thus the price we pay for analytical insight is a complete loss of spatial detail, which may or may not affect the qualitative behaviour of the IPS considered.

Besides mean-field approximations, there is another useful method for exploring IPS dynamics – an option in between the numerical simulation of the original IPS and the analytical discussion of its mean-field approximation (T. Czárán and R. Hoekstra, unpublished results). The idea on which it is based is the following: it is always possible in a stochastic cellular automaton to calculate the probability of any imaginable neighbourhood configuration from the frequencies of the cell states at time t (that is, the state frequency vector $\mathbf{x}(t)$), provided that

1. the grid is infinitely large; and
2. the cell states are randomly reallocated every time unit after updating.

Since it is the neighbourhood configuration at time t that determines the state transition of the focal cell by time $t + 1$ according to the updating rule, and the probability of each possible neighbourhood can be calculated, a state transition matrix $\mathbf{T}(\mathbf{x})$ can be computed for any state frequency vector $\mathbf{x}(t)$. Thus the IPS model is reduced to a non-linear Markov chain. Such approximations might be termed **configuration-field** systems, a detailed example for which is given in Section 5.3.3. The main difference between a mean-field and a configuration-field approximation is that the latter preserves more from the assumptions of the original IPS (U. Dieckman, personal communication). The effect of neighbourhood size is easy to consider in a configuration-field model; it is possible, for example, to calculate the configuration-field approximation with a Moore and a Neumann neighbourhood for the same IPS, and to ask about possible differences in these two cases; this distinction cannot be made in a mean-field model. On the other hand, a configuration-field system offers the possibility of treating the problem in a closed form like a frequency-dependent transition matrix – which does not usually mean analytical tractability· at the same time, of course.

Even very simple conjectures on the behaviour of IPSs might be very awkward, if at all possible, to prove if no mean-field or configuration-field approximation is applicable for a problem (this can happen, for example, if the updating rules depend on spatial direction, as in the model of Section 5.3.7). The analytical treatment of IPSs can be very difficult even if an approximation is straightforward, however. For illustrations of the difficulties and the occasional beauty of rigorous proofs for IPSs, the reader is referred to Liggett (1985). The transition from discrete to continuous time (that is, from synchronous to asynchronous updating) helps somewhat in making certain IPSs analytically tractable. This effect is analogous to that of the transition from difference to differential equations in calculus; therefore it is a bit less unexpected than the methodological benefit from introducing stochasticity into deterministic CA.

5.2.3 Aspects of complexity in interacting particle systems

With simple local rules, IPSs are capable of generating very complicated patterns on the lattice level, sometimes comparable to those produced by natural systems. These features make IPS models ideally suited for the study of a wide spectrum of scientific problems in many disciplines, namely those connected with the collective behaviour of many simple, locally interacting entities. The complexity of IPS dynamics has different aspects. Perhaps the most striking one is that some of the lattice-level properties are not even explicable or interpretable in terms of the local rules; one may need to use different concepts for that. Many CA evolve self-similar patterns, for example, which can be characterized by a certain fractal dimension. Neither self-similarity nor fractal dimensionality is a concept applicable to the state transition rules or the local states of the cells; however, these are **emergent** (Hogeweg, 1988) properties of the pattern produced by the system on a higher level of its organization.

The topological complexity of CA structures is very convincingly demonstrated in the 'atlas of basin of attraction fields' by Wuensche and Lesser (1992), which is a graphical representation of the complete topology of the attractor structure in one-dimensional Boolean CA of length 15 cells with a three-cell neighbourhood. In spite of the very reduced extension of the lattice and the simplicity of the transition rules applied, the complete topology of the CA attractors needs almost 150 pages to be visualized graphically – and no simpler method is known for exploring the attractor structure.

The emergence of complex, organized mesoscale patterns from simple or irregular initial states through stochastic particle interaction rules is a common feature of many IPS. It applies to models of actual physical or biological systems, as well as to **metaphoric models** built with the intention to illustrate physical or biological principles without referring to any actual phenomenon in the 'real' world. The thermodynamic criterion for the emergent feature of **decreasing configuration entropy** is that the interactions of cells must be **dissipative**, which implies that the local updating rules of a self-organizing IPS must be non-linear. This is a necessary (but, of course, not sufficient) condition for the appearance of self-organized emergent structures in stochastic IPS models, whatever (if any) actual scientific problem they are motivated by. The non-linear character of the rules of cell interactions in IPSs is not always as obvious at first sight as it is in the differential or difference equations of dissipative i-state distribution models, however, since the updating rules are not usually formulated in algebraic terms. Note that the non-linearity of the updating rule is not a necessary criterion for deterministic CA to evolve highly ordered and complicated structures from very simple ones, since their configuration entropy is not increased by probabilistic events; therefore it does not need to be compensated by 'energy' dissipation.

IPSs are complex also in the sense that they can mimic the intricate behaviours of manmade and natural objects surprisingly well; albeit in many cases the representation is 'metaphoric', so to say, since the actual mechanisms producing the observed behaviour is difficult or impossible to map in the local rules of the model. Such metaphoric models of complex behaviours might mimic self-replication (Box 5.1) or universal computing abilities as mentioned earlier, but also, for example, some aspects of embryonic development (cf. Ermentrout and Edelstein-Keshet, 1993), which is known to be one of the notoriously difficult problems in biology. From our viewpoint, the most important applications of IPSs are population dynamical models that generate organized patterns of species abundance in space and time. These are typically **not** metaphoric in the above sense: their local rules are set up so that they simulate real mechanisms of interactions in a simplified manner.

5.3 INTERACTING PARTICLE SYSTEMS AND CELLULAR AUTOMATA IN ECOLOGY

The spectrum of existing IPS approaches in ecology ranges from models of the speed and geometry of bacterial colony growth on different substrates and different abiotic milieu conditions (Ben-Jacob et al., 1994a, b), through overgrowth-competition patterns and dynamics in sessile marine invertebrates (Karlson and Jackson, 1981; Karlson, 1984; Karlson and Buss, 1984), local competition and dispersal dynamics among plants (van Tongeren and Prentice, 1986; Crawley and May, 1987; Czárán, 1989; Inghe, 1989; Auld and Coote, 1990; Palmer, 1992; Perry and Gonzalez-Andujar, 1993), explicitly spatial host–parasitoid (Comins et al., 1992) or predator–prey (De Roos et al., 1991; Wilson et al., 1993) interactions of animals, to larger spatiotemporal scale metapopulation or landscape-level approaches (MacKay and Jan, 1984; von Niessen and Blumen, 1986; Green, 1989; Caswell and Etter, 1993; Etter and Caswell, 1993; Liu, 1993; Wissel and Jeltsch, 1993; Dytham, 1994, 1995; Jeltsch and Wissel, 1994; Grimm et al., 1996; Jeltsch et al., 1996). The following is a selection of interesting problems from different fields of theoretical ecology, to which IPS and CA models with different spatial and temporal scales have been applied.

5.3.1 Discrete individuality and dynamical coexistence

In a recent paper, Durrett and Levin (1994a) systematically compared the predictions of four different modelling approaches representing a specific, unusual form of frequency- and density-dependent interaction between two populations. The aim of the study was to demonstrate the effects of differences in fundamental model assumptions on the prospects for coexistence, with the interaction pattern itself kept invariant. The comparison included (i) a non-spatial ODE model, (ii) a reaction–diffusion model, (iii) a patch model and (iv) an IPS.

In the ODE model, the densities of the two populations, H and D are $u(t)$ and $v(t)$ at time t, respectively, which we denote by u and v for simplicity in the sequel. The actual growth rates of the populations consist of

1. their intrinsic growth rates r_u and r_v (these are constants representing the growth rates without the effects of intra- and interspecific interactions);

2. the **frequency-dependent** population interaction terms, $\left(a\dfrac{u}{u+v}+b\dfrac{v}{u+v}\right.$ for H and $\left.c\dfrac{u}{u+v}+d\dfrac{v}{u+v}\right.$ for D$\left.\right)$; and

3. the **density-dependent** component of the death rates $\kappa(u+v)$, which is the same for the two populations.

The system of ODEs constructed from these assumptions is

$$\begin{aligned}\frac{du}{dt}&=u\left\{r_u+a\frac{u}{u+v}+b\frac{v}{u+v}-\kappa(u+v)\right\}\\ \frac{dv}{dt}&=v\left\{r_v+c\frac{u}{u+v}+d\frac{v}{u+v}-\kappa(u+v)\right\}.\end{aligned} \tag{5.1}$$

Note that the frequency-dependent interaction terms do not depend on the actual densities, only on the **frequencies**, that is, on the density shares $u/(u+v)$ and $v/(u+v)$ of the species within the total density. This is equivalent to assuming that the actual **number** of interactions per unit time makes no difference in the instantaneous change of population growth rate; the only relevant factor in this respect is the **ratio** of intra- and interspecific interactions. Such an assumption is rather questionable in terms of ecological reality, albeit it is very common in models of game dynamics, partly because of the mathematical convenience it offers when it comes to the stability analysis of the system. Durrett and Levin (1994b) also give a game theoretical interpretation to the (5.1) model, assuming that the frequency-dependent part of the interaction is the result of a two-strategy game, the payoff matrix of which would be

	H	D
H	a	b
D	c	d

The entry c of the payoff matrix specifies the direction and the strength of the effect of population H on the growth rate of population D. Positive entries represent beneficial effects, negative entries stand for adverse effects in the general sense. The directly density-dependent decline term $\kappa(u + v)$ of (5.1) is the same for the two populations, which is an implicit postulation of the similarity of H and D regarding the exploitation of resources that are **not** affected by the frequency-dependent interaction.

The other three models are all spatial analogues of (5.1). The reaction–diffusion version is (5.1), with the simplest possible 2D Fickian diffusion terms added to both equations (cf. Chapter 3):

$$\begin{aligned}\frac{\partial u}{\partial t} &= \left\{\frac{\partial^2 u}{\partial x^2} + \frac{\partial^2 u}{\partial y^2}\right\} + u\left\{r_u + a\frac{u}{u+v} + b\frac{v}{u+v} - \kappa(u+v)\right\} \\ \frac{\partial v}{\partial t} &= \left\{\frac{\partial^2 v}{\partial x^2} + \frac{\partial^2 v}{\partial y^2}\right\} + v\left\{r_v + c\frac{u}{u+v} + d\frac{v}{u+v} - \kappa(u+v)\right\}.\end{aligned} \quad (5.2)$$

Of course, u and v are functions in both space and time: $u(x, y, t)$, $v(x, y, t)$, in the reaction–diffusion model.

The patchy system is a discrete event simulation model, the space of which consists of 2500 habitat patches, each capable of maintaining more than one individual. The rules of the patch model are the following (Durrett and Levin, 1994b).

1. Migration. Each individual changes its spatial location j (home patch) at a probability μ, and when it moves it moves to a randomly chosen patch. This random dispersion is the only interaction among cells. No neighbourhood relations are relevant in migration.
2. Deaths due to crowding. Each individual on patch j at time t dies at a local density-dependent probability $\kappa\{u_j(t) + v_j(t)\}$, where $u_j(t)$ and $v_j(t)$ are the numbers of H and D individuals, respectively, on patch j at time t.
3. Interaction step. Species H individuals on patch j experience a birth (or death) probability $ap_j(t) + b\{1 - p_j(t)\}$, while D individuals experience a birth (or death) probability $cp_j(t) + d\{1 - p_j(t)\}$, depending on the local frequency $p_j(t) = u_j(t)/\{u_j(t) + v_j(t)\}$ of H individuals on patch j at time t. Whether it is birth or death that actually occurs, depends on the sign of the corresponding interaction term.

The IPS model differs from the patch model in that the habitat patches are arranged in a 50 × 50 square lattice, and a neighbourhood is defined for both the migration step (1) and the game step (3). The neighbourhoods might be different for the two; migrant individuals were allowed to land within the Neumann neighbourhood (four orthogonal nearest neighbours) of the cell they started from in all simulations, whereas an individual was affected in terms of death and reproduction only by the individuals within either its Neumann neighbourhood, or the larger 5×5 square neighbourhood. The IPS rules were equivalent to those of the patch model (1)–(3) in all other respects. Thus the patch model is in fact a spatially randomized version of the IPS on a 50×50 lattice, in which each site represents a habitat patch capable of maintaining more than one individual.

The predictions pertaining to the stability of equilibria in the four models may or may not differ, depending on the structure of the payoff matrix. If the payoffs a, b, c, d are such that (5.1) admits a single global interior equilibrium, that is, if the ODE system evolves towards unconditional coexistence, all the other three systems agree. This is the case when interspecific interactions are of more benefit (or less cost) in terms of changes in the growth rates for both species than the interactions with conspecifics, e.g., if the payoff matrix is

	H	D
H	−0.6	0.9
D	0.9	−0.7

a uniform spatial pattern of coexistence develops in all the corresponding spatial models.

Spatial aspects start playing a role when the payoff matrix defines conditional exclusion of one of the species, the loser depending on the initial densities in the ODE model (Figure 5.3). This comes about if the interaction with conspecifics do more benefit (or less harm) than interaction with the other species, e.g., if

	H	D
H	0.6	−0.9
D	−0.9	0.7

then the ODE system is bistable and so is the patch model. The reaction–diffusion model and the IPS may not agree in this case, however: they both predict unconditional exclusion of one of the species for almost any initial conditions. Which population wins depends on the actual entries of the payoff matrix. Roughly speaking, whichever species has the larger basin of attraction in the ODE representation, will ultimately take over in the reaction–diffusion model and the IPS (the winner would be D for the payoff matrix above). The exclusion of the loser proceeds in a spatially organized

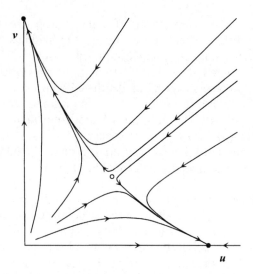

Figure 5.3 Phase portraits of a bistable version of (5.1), leading to the conditional extinction of one species and the persistence of the other. Which species persists depends on the initial densities. Parameters: $a = 0.7$, $b = 0.4$, $c = 0.4$, $d = 0.8$ (after Durrett and Levin (1994a)).

manner, by forming travelling waves of the winner in the reaction–diffusion approach, and expanding clumps of cells monopolized by the winner in IPS. For this effect to appear, 'seeds' of monospecific clumps must be formed by chance, either due to inhomogeneities in the initial pattern, or by the spatial contingencies of the process. Once formed, these origins of exclusion should not be dissolved by mixing the system drastically, otherwise the dynamics of the system degenerates to that of the non-spatial, bistable case. The new dynamical properties of the spatially explicit models (the reaction–diffusion system and the IPS) arise from the joint effect of local interaction and limited dispersion for this specific payoff configuration – if any one of these two factors is missing, the effect does not show up.

The most interesting conclusion comes from the case when the payoff matrix is such that one of the species (D) is a general 'mutualist', the other (H) is a general 'competitor'. This happens to be the case, e.g., for the payoff matrix

	H	D
H	−0.6	0.9
D	−0.9	0.7

H has a clear dominance over D, since D-type individuals feel more intensive competition when playing against Hs, and get less benefit from interactions with Ds, in terms of the coefficients of the frequency-dependent part of their growth rate. The ODE model of this situation predicts extinction for both populations, just like the reaction–diffusion system with a reflecting boundary (see Durrett and Levin (1994b) for a proof). The two species approach the state of global extinction together, since H would die out without D, but it suffers less from the competitive interaction than D when D is rare ($-0.6 > -0.9$), and benefits more from the mutualistic effect when D is common ($0.9 > 0.7$). This keeps the densities of the two populations positive, but steadily declining at all times.

Both discrete systems can admit coexistence with this configuration of the payoffs: the average densities in the patch model settle to an interior quasi-equilibrium, whereas the IPS shows quasi-limit-cycle or quasi-fixed-point behaviour, depending on grid size and interaction neighbourhood size (Figure 5.4). Coexistence is the result of two inseparable factors in both models: one is that the species are represented by a few discrete individuals in each interaction neighbourhood, not by infinite numbers of infinitesimally small ones; the other factor is the resulting demographic stochasticity within the neighbourhoods, which, given the small

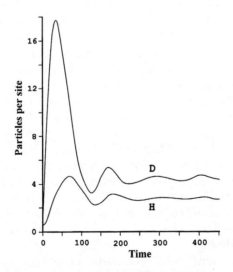

Figure 5.4 The temporal behaviour of the patch model with a mutualistic–competitive payoff structure. Parameters: $a = -0.6$, $b = 0.9$, $c = -0.9$, $d = 0.7$, $\kappa = 0.02$, $\mu = 1.0$, number of patches: 2500, with the 2, 8 initial condition (after Durrett and Levin (1994b)).

local populations, can lead with sufficient probability either to the local extinction of H directly, or to the local extinction of D, and thus the subsequent extinction of H indirectly. Empty cells are then ready to be recolonized by D, which can be regarded as a fugitive species *sensu* (Huffaker, 1958).

The similarity of the patch model with these payoffs, and the predatory metacommunity system of Caswell and Cohen (1991b) (cf. Section 4.2.7) is striking. The metacommunity model can be in fact considered as a version of the patch model (which is itself a spatially randomized version of the IPS), whereby the local 'catastrophes' creating empty patches to be colonized by the prey species result from the internal dynamics of the patches (overpredation and joint extinction as a consequence). Therefore the probability of a local catastrophe occurring is dependent on local frequencies. Such local catastrophes are necessary for both the metacommunity system and the patch model to admit coexistence. These are possible through demographic processes only if the local populations are small – at least some of the time – and the individuals are discrete. Note that any metapopulation model can be reinterpreted this way, if subpopulations are small, prone to local extinction because of demographic stochasticity, for example. Thus the surprising predictive power of metapopulation models might be attributable, at least in part, to their implicit assumption of discrete individuality at the local scale.

The same mechanism works in the IPS model as well, but it is modified by the local character of both the game step and dispersal, which leads to the grid size effect and the neighbourhood size effect as shown on Figure 5.5. The grid size effect of the IPS model is clearly related to the asynchrony of the cyclic dynamics in different parts of the grid. Within small clumps of cells, local synchrony is maintained by local interaction and limited dispersion, but these effects, just because of being local, cannot synchronize distant regions of the grid. The larger the grid, the more pronounced the asynchrony effect, of course, and thus the smaller the amplitude of the oscillations in the regional averages of densities. Increasing the interaction neighbourhood, on the other hand, affects the dynamics through the process of averaging out abundance differences among the cells within the neighbourhood, thus enforcing synchrony on large clumps of cells, which has a clear destabilizing effect on regional dynamics. Note that the interaction neighbourhood consisted only of the focal site in the patch model, which in fact explains in itself why the patch model is more stable than the IPS with a 5×5 neighbourhood.

The most important lesson from this study is that the **existence of individual objects as discrete entities** might be responsible for the coexistence of species that would die out otherwise. This conclusion applies to a number of types of population interaction, assuming different dynamics, possibly more realistic in biological terms, but not to all types. In the above study, discrete individuality proved to be important dynamically

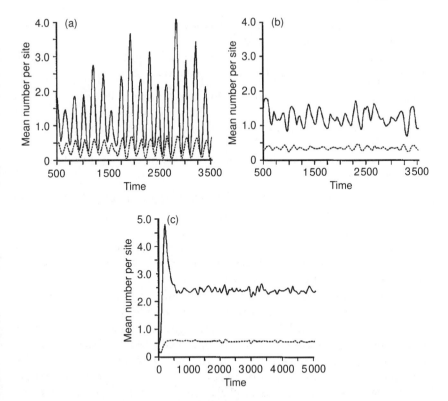

Figure 5.5 The temporal behaviour of the IPS with the same mutualistic–competitive payoff structure. Parameters as in Figure 5.4 and (a) grid size: 50 × 50, neighbourhood size: 5 × 5; (b) grid size: 150 × 150, neighbourhood size: 5 × 5; (c) grid size: 50 × 50, neighbourhood size: four orthogonal neighbours (Neumann neighbourhood) (after Durrett and Levin (1994b)).

only in the case when the competitively superior species was in need of the presence of the subordinated one, so that without the weak species present, even the strong one died out. If the individuals of the dominated species are discrete, the dominating species can exploit them only up to a certain point, namely until the last individual of the dominated species disappears completely from the neighbourhood. Then the dominant species also dies out locally, and makes the site free for the dispersing propagules of the subordinated species to get established. In other words the fugitive species takes advantage of its own local extinction by also driving the dominant species to extinction locally. Discrete individuality plays a central role, essentially for the same reason, in the coexistence of competitive–mutualist replicator molecules in the IPS model of prebiotic evolution by Czárán and Szathmáry (1997), suggesting that obligate mutualism is another type of interaction that might require the populations to be 'particulate' to be able to coexist.

In conclusion, it seems to be the general rule that for discrete individuality to produce coexistence in pairwise population interactions:

1. the interaction of the species should be asymmetric, but the dominant species must be dependent on the subordinated one;
2. the subordinated species must be able take a relative advantage, one way or another, from being rare and
3. the interaction must be local; the dynamics of the localities must be asynchronous.

Points 1 and 2 constrain the types of ecological interactions in which discrete individuality can play a role in coexistence. Besides asymmetric competition, the most plausible interaction type satisfying these assumptions would be, of course, predation. More than twenty years ago, Maynard Smith (1974) published a very simple model that was most probably the very first IPS implementation of a two-species predator–prey metacommunity (although at that time none of the terms 'metacommunity' or 'interacting particle systems' was known; see also Renshaw (1991)). The possible occupancy states of the cells changed from empty through increasing densities of prey and different density combinations of prey and predator back to empty, eight different states altogether. The prey died out locally because of overexploitation by the predator, whereas the predator became extinct from starvation in the absence of prey. The transitions from one discrete occupancy state to the other were probabilistic, invasion probabilities of both prey and predator depending on the states of the focal cell and the cells within its Neumann neighbourhood. The model was very stably coexistent, albeit any single cell went through the empty state regularly. The recurrence of regionally asynchronous local extinctions is again a necessary requirement for regional coexistence in this model. Implicitly it is the consequence of assuming discrete individuality, since if the cells would behave asymptotically, complete extinction would never occur, and recolonization by the prey would be impossible. Soon after Maynard Smith, Zeigler (1977) used a similar discrete-event simulation model to clarify the effect of varying the dispersal rates of both the prey and the predator.

5.3.2 Interacting particle system models of competing metapopulations with temporary and permanent habitat destruction

Recently Caswell and Etter (1993) constructed an IPS model of two competing metapopulations, one dominant, the other fugitive, modifying the mean-field metacommunity model of Caswell and Cohen (1991b; Section 4.2.10) to include local interactions and limited dispersal. The novelty of this model compared to the mean-field version is that the state transition probabilities depend on the occupancy states of the sites in the actual neighbourhood, instead of the average occupancies within the grid. The authors also introduced an ecological effect of intermediate spatial scale, one between the scale of a single site and that of the grid, by allowing for the spatial clumping of local catastrophes. This made the IPS results different from what was predicted by the mean-field model, because the empty clumps of sites created by the catastrophes could be colonized by the fugitive species only gradually, starting from the periphery of the clumps, proceeding inwards. The effective colonization rate of the weak species was therefore less than it could be with the same number of spatially scattered spots of local catastrophes, which in turn diminished the parameter domain where coexistence was possible.

Dytham (1994, 1995) studied an IPS version of the spatially implicit competitive metapopulation model of Nee and May (1992) and the related spatially semi-explicit system of Hanski (1994a) (Section 4.2.12). Competition was also one-sided in the IPS model, and the fugitive species was a better disperser than the dominant. Colonizations were strictly local, affecting only the four orthogonal nearest neighbours of the donor patch (in Nee and May (1992) any empty patch could be colonized from any occupied one with the same probability; in Hanski (1994a) the biotic isolation between two patches depended on the distance between them). The question was the same: how does the permanent elimination of a fraction $1 - h$ of the patches from the grid affect the persistence and the coexistence of the two species? The results of the IPS model were slightly different from the other two: coexistence of the two species was possible even if none of

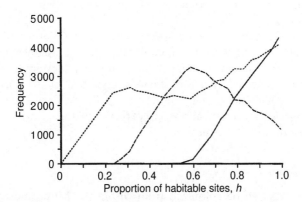

Figure 5.6 Number of sites occupied by the superior competitor (solid line, extinction rate $e = 0.1$, colonization rate $c = 0.2$), the inferior competitor (dashed line, $e = 0.1$, $c = 0.5$) and empty, but still available for colonization (dotted line) in a 100×100 grid, when a proportion $1 - h$ of the sites are permanently destroyed (after Dytham (1995)).

the patches was eliminated, but a moderate level of habitat destruction helped the fugitive species to take over. Strong destruction drove even the more dispersive species to extinction, because its effective colonization rate decreased below the extinction rate due to the scarcity of neighbouring patches (Figure 5.6).

5.3.3 The temporal refuge effect in one-sided competition; an example for a configuration-field approximation

High spatial dispersivity seems to be generally considered as an essential part of the fugitive strategy in habitats where asynchronous local catastrophes are frequent. The IPS model of Caswell and Etter (1993) is an example also for this, demonstrating that a competitively weak, but efficiently dispersing species can persist for a long time in spite of the presence of a competitively dominant, but slowly dispersing one, because it can quickly colonize empty patches produced by local catastrophes; it can develop to maturity and send out new dispersers before being abolished by the superior competitor. In fact the essential part of this strategy is that the fugitive species be able to produce fertile offspring faster than its competitor – how it can achieve this, makes no difference in the outcome. An alternative of the refuge effect due to high spatial dispersivity could be that the two species disperse equally well, but the weak competitor behaves like an r-strategist: it grows faster while the habitat is unsaturated, but loses competition if the local environment remains the same for a long time. This situation can be conveniently modelled by a simple two-species nine-state metacommunity IPS (T. Czárán and R. Hoekstra, unpublished results) for competing yeast populations, the state transitions of which are discretized and simplified representations of the colonization–competition dynamics depicted on Figure 5.7.

The arena of two competing species is a 50 × 50 lattice of sites. The possible states of the sites are combinations of abundance for the two species, h and i; E means the 'empty site' state, small letters (h, i) stand for few, capital letters (H, I) denote many individuals of the corresponding species present on a site. There are three possible types of site state transition in the model: (i) colonizations, (ii) extinctions due to local catastrophes and (iii) competitive transitions.

Colonizations

Empty (E) sites can be colonized from within the Moore neighbourhood centred on them, but only source sites harbouring many individuals (that is, sites in any of the states H0, Hi, hI, 0I or HI) send out colonizers. The simple transition rule of colonization is that H× sites send 'h' type colonists and ×I sites send 'i' type ones, in each case with a dispersal probability d onto each of the empty (E) sites around them. h0 and H0

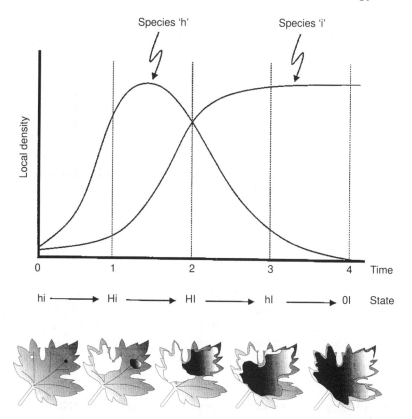

Figure 5.7 Hypothetical local dynamics on a single site invaded by both species. Abundance states: hi, few 'h' and few 'i'; Hi, many 'h' and few 'i'; HI, many 'h' and many 'i'; hI, few 'h' and many 'i'; 0I, no 'h' and many 'i'.

sites can be overcolonized by species 'i', but 0I and 0i sites cannot be invaded by 'h'. Different colonization events directed to a certain empty site are independent of each other in the sense that the success of the colonization of one species is not influenced by the colonization of the other. For example, if there are three source sites, two hI and one HI in the neighbourhood of an E site, the probability that the empty site remains in state E is $(1 - d)^4$; the chance that it will be colonized by species 'h' alone or by species 'i' alone is $d(1 - d)^3$ and $(1 - d)[1 - (1 - d)^3]$, respectively; the probability that the site will be colonized by both species so that it turns to state hi is $d[1 - (1 - d)^3]$. Equal dispersivity is inherent in the assumption that d is the same for both species. Note that the dispersal parameter is a spatial equivalent of the colonization rate of metapopulation models (see Chapter 4).

176 Site-based neighbourhood models

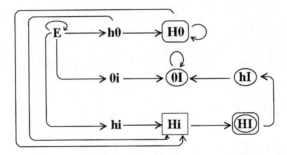

Figure 5.8 State transition routes for the competition IPS. Filled arrows: colonization events; open arrows: local state transitions; square-boxed states: 'h' colonizers; ellipse-boxed states: 'i' colonizers.

Local extinctions

Any site in a state different from E will be converted to E with a probability e, which is the chance of a local catastrophe to occur per site. This parameter is called the extinction probability; it is essentially the same as the extinction parameter e in metapopulation models.

Competitive transitions

Colonizations and extinctions are stochastic events, but if neither of them occurs on a site, that site will either stay as it was (H0 and 0I) or it steps forward to its next state deterministically, according to the order specified in Figure 5.8, showing the permissible state transition routes for colonization and competition. Note that 0I is a sink state of the graph: all possible routes with species 'i' present converge there: species 'i' is an absolutely superior competitor over species 'h'. In other words the latter has no chance of surviving in an undisturbed habitat.

The model has two parameters altogether: d, the dispersal parameter and e, the rate of extinction – both range from zero to one. Figure 5.9 shows the results of scanning the parameter space with steps of 0.1 in both dimensions. The initial pattern was randomly constructed by dispersing 250 propagules of species 'h' and another 250 of 'i' within the grid – this corresponds to a 10% initial saturation level for each species. In spite of the competitive dominance of species 'i', the system produces all the possible outcomes, including stable coexistence, both cases of regional competitive exclusion of one species by the other, and the regional extinction of both species, depending on the actual value of e and d. For small values of dispersal (colonization) and large extinction rates, the model predicts collective regional extinction, just as would any metapopulation model. Small extinction rates allow the sites to persist long enough for the dominant competitor to take over everywhere and drive the other species to regional extinction. The range in which species 'h' survives and 'i' dies out is where dispersal is high, and the rate of local extinction is moderately high (0.6–0.8). This range represents 15% of the parameter space. The domain of coexistence is characterized by medium values of the extinction rate and medium to high values of dispersal; it takes about 23% of the parameter space.

Different coexistent parameter combinations produce rather different spatial patterns (Figure 5.10), suggesting that the emergent spatial configurations might play a role in shaping the dynamics of the system. A semi-analytical configuration-field approximation can be constructed to check this, which omits the effect of any mesoscale pattern arising in the IPS, but gives more general results than numerical simulation alone. The only new assumptions of this model compared to those of the IPS are that (i) the grid is infinitely large, and (ii) the site states are randomly reallocated in each time unit before colonizations start. Spatial randomization enables the probability of any possible neighbourhood configuration, and thus the probability of any colonization event, to be directly calculated by combinatorial reasoning from the actual site state frequency vector

$$\mathbf{X}(t) = \{x_{\mathrm{E}}, x_{\mathrm{h0}}, x_{\mathrm{0i}}, x_{\mathrm{hi}}, x_{\mathrm{Hi}}, x_{\mathrm{HI}}, x_{\mathrm{hI}}, x_{\mathrm{0I}}, x_{\mathrm{H0}}\}.$$

Colonization rate \ Extinction rate	0.0	0.1	0.2	0.3	0.4	0.5	0.6	0.7	0.8	0.9	1.0
0.0	B	O	O	O	O	O	O	O	O	O	O
0.1	I	I	I	I	O	O	O	O	O	O	O
0.2	I	I	I	I	I	B	O	O	O	O	O
0.3	I	I	I	I	I	B	O	O	O	O	O
0.4	I	I	I	I	I	B	H	O	O	O	O
0.5	I	I	I	B	B	B	H	H	O	O	O
0.6	I	I	I	B	B	B	H	H	O	O	O
0.7	I	I	B	B	B	B	H	H	O	O	O
0.8	I	I	B	B	B	B	H	H	O	O	O
0.9	I	I	B	B	B	B	H	H	H	O	O
1.0	I	I	B	B	B	B	H	H	H	O	O

Figure 5.9 The outcomes of competition in the IPS model after 10000 generations for combinations of the rate of local extinction (*e*) and dispersivity (*d*). 0: both species die out; I: species 'h' dies out; H: species 'i' dies out; B: both species survive.

178 Site-based neighbourhood models

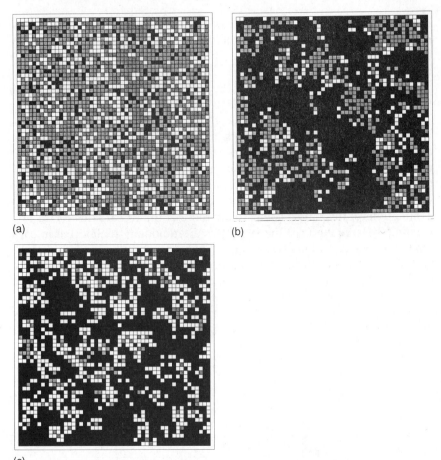

Figure 5.10 Patterns of coexistence for some parameter combinations in the IPS model after 50 generations. Initial densities were 0.1 for both 'h' and 'i' propagules. (a) $e = 0.2, d = 0.7$; (b) $e = 0.5, d = 0.4$; (c) $e = 0.5, d = 1.0$. Black: empty; dark grey: only 'h'; white: only 'i'; light grey: both.

Using the colonization probabilities thus calculated and the extinction parameter, a non-linear Markov chain model can be constructed (see Box 5.2), the results of which can be compared to the outcomes of the IPS.

Figure 5.11 shows the results of the configuration-field approximation, using the same parameter set as for the IPS. It is obvious that spatial randomization does not change the qualitative behaviour of the model: the domains of both kinds of competitive exclusion, joint extinction and coexistence, are all represented in the parameter space, but their sizes are different from those predicted by the IPS version. The most counter-intuitive result is that the coexistent domain has increased in the configuration-field model to

IPSs and CA in ecology 179

Extinction rate

Colonization rate	0.0	0.1	0.2	0.3	0.4	0.5	0.6	0.7	0.8	0.9	1.0
0.0	B	O	O	O	O	O	O	O	O	O	O
0.1	I	I	I	I	I	O	O	O	O	O	O
0.2	I	I	I	I	I	I	B	O	O	O	O
0.3	I	I	I	I	I	B	B	B	O	O	O
0.4	I	I	I	B	B	B	B	B	O	O	O
0.5	I	I	I	B	B	B	B	B	O	O	O
0.6	I	I	B	B	B	B	B	B	H	O	O
0.7	I	I	B	B	B	B	B	B	H	O	O
0.8	I	I	B	B	B	B	B	B	H	O	O
0.9	I	B	B	B	B	B	B	B	H	O	O
1.0	I	B	B	B	B	B	B	B	H	O	O

Figure 5.11 The outcomes of competition in the configuration-field model, for combinations of the rate of local extinction (e) and dispersivity (d). Key as in Figure 5.9.

almost twice its size in the IPS: it occupies 42% of the parameter space. This happens mainly at the expense of the part where 'h' wins in the IPS: the 'h wins' interval decreased to 4.5%, suggesting that it is the competitively inferior species 'h' which benefits more from spatial aggregation. The loss from the range where 'i' wins competition is much less.

According to the results of both the IPS and the corresponding configuration-field model, the coexistence of the dominant and the subordinate species, and even the competitive exclusion of the dominant proved to be possible in a non-equilibrium situation. The subordinated species need not be the better disperser for this – the only temporary advantage compensating for its absolute competitive inferiority is its higher initial rate of population growth. The subordinated species can also benefit more from spatial constraints (that is, local interaction and limited dispersal) than the dominant.

The extensive numerical investigation of the matrix model suggests that the configuration-field system has a single, globally stable equilibrium point for any parameter combination. As far as one can judge from the outcomes of the IPS model, this characteristic remains unchanged if the system is allocated to topographical space. The mesoscale pattern arising from the limited dispersion of the species in the IPS model does not change the predictions in the qualitative sense, but it helps the inferior competitor in quantitative terms.

Box 5.2 The configuration-field approximation

The site state frequency vector is

$$\mathbf{X}(t) = \{x_E, x_{h0}, x_{0i}, x_{hi}, x_{Hi}, x_{HI}, x_{hI}, x_{0I}, x_{H0}\}.$$

Extinction and competition are both neighbourhood-independent; only the colonization probabilities depend on the structure of the neighbourhood. The nine different abundance states of the model can be classified into four categories according to the type of propagules produced: sites in the E, h0, 0i, and hi states cannot colonize, H0 and Hi sites can emit 'h', 0I and hI sites can send out 'i' type propagules, and HI sites can produce colonizers of both species. The fraction of non-colonizing sites in the grid at time t is $p(t) = x_E + x_{h0} + x_{0i} + x_{hi}$; that of 'h' and 'I' colonizers is $q(t) = x_{H0} + x_{Hi}$ and $r(t) = x_{0I} + x_{hI}$, respectively; the fraction of sites capable of colonization by both species is $s(t) = x_{HI}$. p, q, r and s are also the probabilities that a randomly chosen site belongs to the corresponding category. The probability that a randomly assembled neighbourhood consists of exactly k_0, k_h, k_i and k_{hi} sites of the corresponding colonizer type is

$$P_{k_0,k_h,k_i,k_{hi}} = \frac{8!}{k_0! k_h! k_i! k_{hi}!} p^{k_0} q^{k_h} r^{k_i} s^{k_{hi}}. \quad [5.1]$$

Of course, $k_0 + k_h + k_i + k_{hi} = 8$ is the size of the neighbourhood, with the focal site excluded.

Given the (k_0, k_h, k_i, k_{hi}) neighbourhood configuration, the probabilities of no colonization at all ($Q^0_{k_0,k_h,k_i,k_{hi}}$), colonization by 'h' only ($Q^h_{k_0,k_h,k_i,k_{hi}}$) by 'i' only ($Q^i_{k_0,k_h,k_i,k_{hi}}$), and colonization by both species ($Q^{hi}_{k_0,k_h,k_i,k_{hi}}$) on the focal site can be calculated as

$$\begin{aligned}
Q^0_{k_0,k_h,k_i,k_{hi}} &= (1-d)^{(k_h+k_i+2k_{hi})} \\
Q^h_{k_0,k_h,k_i,k_{hi}} &= \left[1-(1-d)^{(k_h+k_{hi})}\right]\left[(1-d)^{(k_i+k_{hi})}\right] \\
Q^i_{k_0,k_h,k_i,k_{hi}} &= \left[(1-d)^{(k_h+k_{hi})}\right]\left[1-(1-d)^{(k_i+k_{hi})}\right] \\
Q^{hi}_{k_0,k_h,k_i,k_{hi}} &= \left[1-(1-d)^{(k_h+k_{hi})}\right]\left[1-(1-d)^{(k_i+k_{hi})}\right]
\end{aligned} \quad [5.2]$$

having recalled the assumption that the immigration events onto an empty site are independent of each other. Thus the probability that a site is surrounded by a (k_0, k_h, k_i, k_{hi}) neighbourhood **and** it is not invaded at all is $c^0_{k_0,k_h,k_i,k_{hi}} = P_{k_0,k_h,k_i,k_{hi}} \cdot Q^0_{k_0,k_h,k_i,k_{hi}}$). The conditional probabilities of the other three different types of colonization events within that neighbourhood can be calculated in the same way, replacing $Q^0_{k_0,k_h,k_i,k_{hi}}$ in turn by $Q^h_{k_0,k_h,k_i,k_{hi}}$, $Q^i_{k_0,k_h,k_i,k_{hi}}$ and $Q^{hi}_{k_0,k_h,k_i,k_{hi}}$. Summing the conditional colonization probabilities for all possible neighbourhood types, the corresponding total colonization probabilities of empty sites are

$$C_0 = \sum_{k_0+k_h+k_i+k_{hi}=8} Q^0_{k_0,k_h,k_i,k_{hi}} \cdot P_{k_0,k_h,k_i,k_{hi}}$$
$$C_h = \sum_{k_0+k_h+k_i+k_{hi}=8} Q^h_{k_0,k_h,k_i,k_{hi}} \cdot P_{k_0,k_h,k_i,k_{hi}}$$
$$C_i = \sum_{k_0+k_h+k_i+k_{hi}=8} Q^i_{k_0,k_h,k_i,k_{hi}} \cdot P_{k_0,k_h,k_i,k_{hi}}$$
$$C_{hi} = \sum_{k_0+k_h+k_i+k_{hi}=8} Q^{hi}_{k_0,k_h,k_i,k_{hi}} \cdot P_{k_0,k_h,k_i,k_{hi}}$$

[5.3]

These are the expected probabilities of no colonization, colonization by 'h', by 'i', and by both species, respectively, of a randomly chosen empty site. Notice that the total colonization probabilities can be calculated at any time t from the actual $\mathbf{X}(t)$ vector of site state frequencies and d, the probability of propagule dispersion.

The rate of the overcolonization of h0 and H0 sites by 'i' propagules – as a result of which they become hi or Hi sites, respectively – is $(C_i + C_{hi})$.

Under these assumptions, the temporal transitions of the state vector $\mathbf{X}$ are governed by the non-linear Markov matrix

$$\mathbf{M}[\mathbf{X}(t)] = \begin{bmatrix} C_0 & e & e & e & e & e & e & e & e \\ C_h & 0 & 0 & 0 & 0 & 0 & 0 & 0 & 0 \\ C_i & 0 & 0 & 0 & 0 & 0 & 0 & 0 & 0 \\ C_{hi} & 0 & 0 & 0 & 0 & 0 & 0 & 0 & 0 \\ 0 & A & 0 & 1-e & 0 & 0 & 0 & 0 & A \\ 0 & 0 & 0 & 0 & 1-e & 0 & 0 & 0 & 0 \\ 0 & 0 & 0 & 0 & 0 & 1-e & 0 & 0 & 0 \\ 0 & 0 & 1-e & 0 & 0 & 0 & 1-e & 1-e & 0 \\ 0 & B & 0 & 0 & 0 & 0 & 0 & 0 & B \end{bmatrix}$$

$A = (1-e)(C_i + C_{hi})$ is the probability of the overcolonization of h0 and H0 sites by 'i', provided that the resident species 'h' does not go extinct in the meantime. $B = (1-e)(1 - C_i - C_{hi})$ is the probability that these sites will be in state H0 by the next generation, that is, neither the local extinction of 'h' nor overcolonization by 'i' occurs. One step of the transition represents one generation:

$$\mathbf{X}(t+1) = \mathbf{M}[\mathbf{X}(t)] \cdot \mathbf{X}(t) \qquad [5.4]$$

Equation [5.4] can be numerically iterated and its stationary state $\hat{\mathbf{X}}$ found.

5.3.4 The role of mesoscale patterns in the dynamics of predator–prey cellular automata

Motivated by a spatial parasitoid–host interaction problem, Comins *et al.* (1992) have built a simple 10-state cellular automaton model, which is actually quite similar to the predator–prey IPS of Maynard Smith (1974) (cf. Section 5.3.1), except that the interaction rules they applied were deterministic. The sites cycled in a predetermined order from empty (A) through prey only with increasing prey density (B, C, D), prey and predator with increasing predator density (E, F, G, H, I), and predator only (J) back to empty (A) (Figure 5.12). The state transitions from empty to few prey (A→B) and from many prey to many prey and few predators (D→E) represent invasions, the former by prey onto empty sites, the latter by predator onto prey-only sites. Invasions are conditional on neighbourhood configuration, but all other state transitions are independent of it, representing mere quantitative changes in local population densities. The invasion conditions are such that if there is no site around of a certain type, capable of emitting colonists, empty (A) sites remain empty and prey-only (D) sites remain prey only; invasion occurs if at least one site of the appropriate 'propagule donor' type is present within the neighbourhood. The condition for prey invasion (A→B transition) was the presence of a B site in the neighbourhood of an empty one in every simulation, whereas predator invasion (D→E transition) was possible with either an E or an F site around a D site.

In terms of biological reality, it is hard to argue in favour of the assumptions that predator invasion is not possible from G, H, I and J sites, if it is from E or F, and that prey cannot colonize empty sites from C and D sites, if it can from B. (One would expect that, with increasing densities, intraspecific density stress would drive more individuals to disperse and search for less crowded habitats nearby. If B sites emit invading prey individuals to A sites, then at least C and D sites should do the same – not counting sites with both prey and predator present; *mutatis mutandis*, the same applies to predator dispersal as well.) The authors of the model are aware of this oddity, so they are cautious with the biological conclusions they draw from the results of this CA model, which was actually designed to

Figure 5.12 Typical site-state patterns of the predator–prey CA. Black sites are empty (state A); grey shades get lighter from state B (few prey) to state J (predator only, white). (a) Rotating spirals: prey invade from B, predators invade from E sites, Neumann invasion neighbourhood. (b) Spatial chaos: prey invade from B, predators invade from F sites, Neumann invasion neighbourhood. (c) Crystalline structure: prey invade from B, predators invade from E sites, Moore invasion neighbourhood (after Comins *et al.* (1992)).

imitate the dynamics of a more realistic, explicit two-species coupled map lattice (CML) system for spatial parasitoid–host interaction.

Starting from a random initial pattern of the 10 states within a sufficiently large (40 × 40) grid, the predator and the prey proved to be coexistent on the grid level, the spatial pattern of coexistence depending on the invasion mechanism. Invasion parameters are implicit in the invasion rules and the type of neighbourhood applied: if a population is a poor invader, an empty site needs a donor site that is densely populated and close if it is to be

184   Site-based neighbourhood models

colonized, whereas a good colonizer population can invade from a scarcely populated and more distant site. Thus the predator invasion condition E, as compared to F, represents larger predator invasiveness, since an E site (with fewer predators than on F sites) within the neighbourhood is sufficient to colonize prey-only (D) sites. Assuming a Moore-type invasion neighbourhood, instead of a Neumann type, means increasing the invasive mobility of both species, since more distant diagonally positioned sites can also send colonizers to empty ones, and this effect is the same for prey and predator.

1. With medium mobility of both populations (prey invasion takes place from few-prey A sites, predator invades from few-predator E sites, but only from the Neumann neighbourhood, which contains the four closest neighbours), coexistence occurred in a spiral wave pattern, each spiral rotating clockwise or anticlockwise, the direction depending on the contingencies of the initial pattern. Spiral waves are a peculiar type of mesoscale spatiotemporal pattern produced by rather different types of spatially explicit models from reaction–diffusion models (Cronhjort and Blomberg, 1994) through coupled map lattices (Comins *et al.*, 1992; Hassell *et al.*, 1994; Rohani and Miramontes, 1995) to IPSs (Boerlijst and Hogeweg, 1991). The common assumption behind all the models producing spiral waves is a cyclic (intransitive) pattern of local interactions among a reasonably large number of states. The intransitive cycle of predator–prey occupancy states in this model mimics the limit cycle behaviour often observed in predator–prey systems, but the completion of the cycle is contingent on neighbourhood effects at the colonization phases. The spiral waves are driven by the spatial coupling of such cycling sites, if the invasive mobilities of the species are close to each other, and both are moderate.
2. If predator mobility is smaller (predator invades only from F sites, which carry more predators than E sites), but all other parameters are the same as in the spiral wave case, the spirals disappear; the irregularly moving fronts of prey-only regions are slowly traced by predator fronts. This irregular spatiotemporal pattern is called 'spatial chaos' by Comins *et al.* (1992).
3. If both prey and predator are good colonizers (with the rules being the same as for the spiral wave case, but applying the Moore-type invasion neighbourhood with all eight neighbours acting as potential donor sites) the system converges to a less dynamic spatiotemporal pattern called a 'crystalline structure' by Comins *et al.* (1992). This structure is not static in the sense that each site goes through all the possible states A–J regularly, but it is static in that the same pattern reappears in every 10th generation (which is the period length of a complete site cycle). Neither moving fronts nor rotating spiral waves can be observed under these conditions.

These three different spatiotemporal patterns represent three different mechanisms for coexistence. One thing is obviously common in each, however: they require heterogeneity in the initial pattern – that is, local asynchrony on the site level – to persist, otherwise the whole grid gets stuck in the prey-only state D (if no predators were present in the initial pattern) or the empty state A (if both populations were present initially). Local asynchrony plays an important role in maintaining coexistence. This effect would be ruined by considering larger invasion neighbourhoods (which mimics even more invasive mobility for both populations), through the synchronization of the grid. But once invasions are localized so that they occur from immediate neighbourhoods only, and the initial state is heterogeneous, the grid remains sufficiently asynchronous to maintain the possibility of recurrent local invasions.

Both the spiral waves and the crystalline structures are the spatial manifestations of the cyclic local dynamics assumed, the difference between them possibly being the actual size of the spirals. With higher invasive mobility for both species, the spirals might degenerate to be the size of a single site; the result is the crystalline pattern. If the mobilities differ markedly in favour of the prey, fugitive coexistence is the mechanism leading to spatial chaos: the prey, being faster in invasion, can colonize empty sites and remain one step ahead of the predator in space. The prey-only wavefront is chased by the predator front relatively slowly. In this case, the spatial contingencies of the initial pattern do not dissipate from the system; that is, no obvious spatial or temporal periodicity arises.

Apart from justified doubts concerning the biological realism of this model, the question arises whether stochastic rather than deterministic updating rules, that is, an IPS model instead of the present CA, would yield the same self-organized structures (spiral waves and crystalline patterns). The IPS model of hypercyclic replicator dynamics (Boerlijst and Hogeweg, 1992) – which is homologous to a community dynamical model with an intransitive cycle of facultative mutualism among nine species – shows that spiral waves might emerge in stochastic systems as well, if the number of populations in the cycle is sufficiently large. It remains an open question anyway, however, whether such spectacular self-organized mesoscale structures in fact play a role in stabilizing real ecological interactions, since no actual field studies indicate the prolonged existence of, for example, rotating spiral waves in any ecosystem.

5.3.5 Plant competition along an environmental gradient

The self-organized formation of heterogeneous community patterns in spatially homogeneous environments is a process that maintains, amplifies and structures initial abundance heterogeneities, whatever mechanism by which these are created. Certain spatial constraints on the dynamics of the populations – the minimum of which are local interaction and short-range dispersion – are necessary to assume in all models capable of producing such self-organized spatial structures. Reaction–diffusion models, coupled map lattices and neighbourhood models share these assumptions, and all of them have ecological applications with self-organized dynamics, in the form of standing, travelling or rotating (spiral) waves of abundance. The range of interaction types producing such persistent spatiotemporal heterogeneities in homogeneous environments is limited, however, albeit spatially inhomogeneous population and community patterns are the rule rather than the exception in nature. In other words, one needs to invoke the inhomogeneity of the environment to explain the observed heterogeneity of community patterns in most cases.

Environmental heterogeneity may account for mosaic-like patterns of abundance in itself, without relying on other spatial constraints of the dynamics, but it can be much more effective acting together with local interaction and limited dispersion. Consider, for example, two similar competing plant populations along a slight environmental gradient. Let the two species differ in a single dynamical parameter (fecundity or death probability, intra- or interspecific competition parameter, etc.) that responds to the environmental gradient. The different responses can be represented by Gaussian-shaped response curves, the optimum points of which do not coincide (Figure 5.13). At the place on the gradient where the response curves intersect, the competitive dominance relation of the two species flips over, splitting the habitat in two regions, each potentially dominated by the species that performs better there. If the response curves are very similar, that is, if they are flat and their optima are close, this change of dominance can be realized in the abundance pattern only if the two species disperse locally, otherwise the mixing effect of long-range dispersal masks the weak

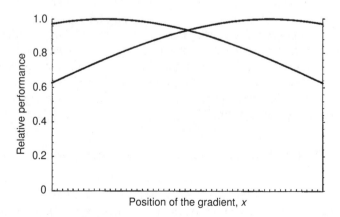

Figure 5.13 Response curves of two hypothetical plant species competing along an environmental gradient (after Czárán (1989)).

IPSs and CA in ecology   187

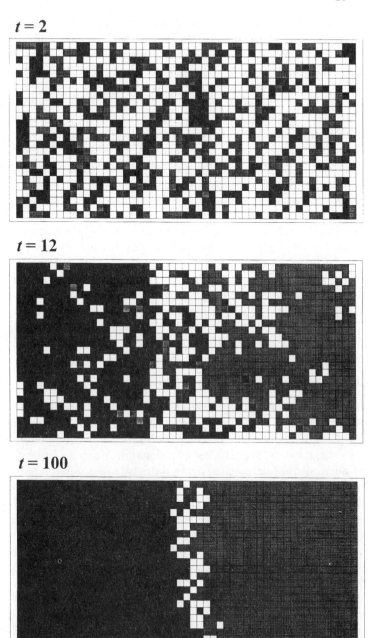

Figure 5.14 Phases of the local segregation process of two competing species with response curves as in Figure 5.13. Black and grey sites are occupied by one species; the two species co-occur in white sites (after Czárán (1989)).

tendency of the two populations to segregate regionally. Assuming a strictly localized dispersion mechanism allows for the appearance of a sharp boundary between the two regions of dominance, even if the response difference between the species is very slight (Figure 5.14., cf. Czárán (1989)). In other words environmental inhomogeneity and spatial dynamics mutually reinforce the inherent tendency of both to induce and to maintain spatial segregation.

An interesting practical corollary of this result is that we might expect to find mosaic patterns in the field even if the inducing environmental inhomogeneity is gradual and so slight that it cannot be measured at all, in which case the long-lasting spatial segregation of two species in the form of a static mosaic pattern seems a mystery at the first sight. Looking at the problem from the other side, even obvious environmental heterogeneities might not show up in the pattern of a plant community, if its species are not very responsive to the heterogeneous environmental factor, and propagules disperse to relatively large distances compared to the scale of the environmental heterogeneity. Thus the classical debate on the indicative value of mosaic patterns in vegetation with regard to environmental heterogeneity can be settled for specific cases at best, and only if the dynamics of the specific situation is known in sufficient depth (Juhász-Nagy, 1986).

5.3.6 Plant competition in a fractal environment

Environmental inhomogeneity can be characterized by different spatial distributions, from a 'smooth' gradual change through the habitat to abruptly changing mosaic-like patchy patterns. If the environmental pattern is mosaic-like, it can be more or less correlated in space, the environmental state of adjacent mosaic patches being more or less similar. These differences in the spatial distribution of environmental factors may have a significant dynamical effect on population interactions, especially among species experiencing the heterogeneity of the environment on different spatial scales. This scale corresponds to the 'area of environmental resolution', over which an individual of the given species can average environmental differences by physiological means. The resolution area of mobile species would be hard to determine, but it is relatively straightforward to do for sessile organisms like terrestrial plants, the root system and the foliage (the environmental 'sensors') of which have a rather definite and fairly static size and shape compared to the home range of animals, for example. A convenient definition of the resolution area for vascular plants would be the ground projection of the root system (if the relevant environmental effect is experienced by the roots: soil humidity, nutrient concentrations, etc.) or that of the foliage (if it is the crown that experiences the environmental effect in question: light intensity, carbon dioxide concentration, air humidity, etc.).

The problem of the connection between the environmental pattern and the dynamics of competing plant populations is also conveniently approached by IPS modelling. T. Kis, I. Scheuring and T. Czárán (unpublished results) recently modified the simulation models of Czárán (1989) and Palmer (1992) to study the dynamical effects of resolution area in different types of environmental patterns.

Assume that the common habitat of s annual plant species consists of a square grid of sites. The sites are equivalent environmentally with respect to all but one environmental factor. The spatial distribution of this factor E within the grid is varied from (i) completely uniform through (ii) gradually changing and (iii) spatially correlated quasi-random to (iv) spatially uncorrelated random. Following the method described by Palmer (1992), the frequency distribution and the spatial distribution of the environmental variable within the grid is controlled by two parameters. Parameter d specifies the range of the environmental variation, d being the standard deviation of the frequency distribution of E over the grid. The mean of the frequency distribution is kept at the same value (zero, for convenience) in all cases, to avoid effects other than those of the range of variation and the spatial configuration of E. Of course, $d = 0$ represents no environmental variation, that is, perfect environmental homogeneity over the grid. The other parameter determines the spatial distribution of E, that is, the spatial correlation of sites with respect to E; this is measured by D, the fractal dimension of the spatial pattern of E. D ranges from 2.0 to 3.0, smaller values representing strong spatial autocorrelation of the environmental variable, that is, smooth change of E across the grid, whereas higher D values indicate less spatial dependence, with adjacent sites weakly correlated (spatial autocorrelation of E approaches zero close to $D = 3.0$). $D = 2.0$ means a linear change of E across the grid, so that the 'landscape' of the environmental variable forms a two-dimensional linear surface (plane), the slope of which depends on d. The direction of the slope is random. With $d > 0$ fixed and D increased, the landscape of E first becomes 'hilly' and then more fine-grained gradually, up to $D = 3.0$, where the values of E are assigned to the sites at random (Figure 5.15). To ensure comparability between environmental patterns of different fractal dimensions, the same frequency distribution of E is preserved for all values of D. The algorithm producing the environmental landscape with a given fractal dimension and frequency distribution is the midpoint displacement procedure with successive random additions, as described in detail by Saupe (1988).

The demography of species i is defined by the expected number b_i of viable seeds produced per plant, the local carrying capacity k of a site, and the dispersal probability p_i. k is the number of adult plants one site can support, regardless of species identity; p_i determines what fraction of the seeds are dispersed over the boundaries of the site. Adults who have dispersed their seeds die. Competition affects seedling mortality, so that if a site gets more seeds than the local carrying capacity k, the surplus is eliminated by an iterative procedure as follows. Let the number of species i seeds within site m be n_{im}. If $\sum_{i=1}^{s} n_{im} > k$ then let $c_{im} = n_{im} \sum_{j=1}^{s} a_{jm} n_{jm}$ be the share of risk for a species i seed among all seeds present in site m to be the one eliminated by the competitive effect of the others. More precisely,

$P_{im} = c_{im} \bigg/ \sum_{j=1}^{s} c_{jm}$ are the probabilities of the mutually exclusive events that n_{im} is decreased by one while all other seed numbers n_{jm} ($j \neq i$) remain the same in a single cycle of the competition algorithm. Iteration with the

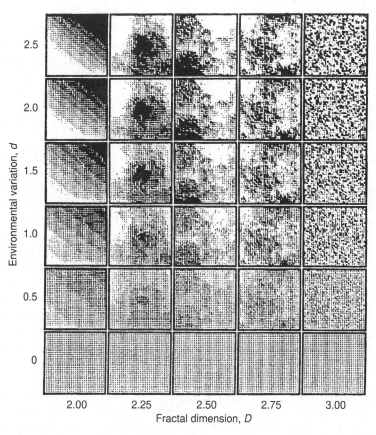

Figure 5.15 Simulated spatial patterns of the environmental variable for different combinations of environmental variation (d) and fractal dimension (D) (after Palmer (1992)).

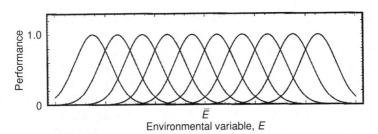

Figure 5.16 Response curves for 10 species along the environmental axis. $\bar{E}$ is the lattice average of the environmental variable.

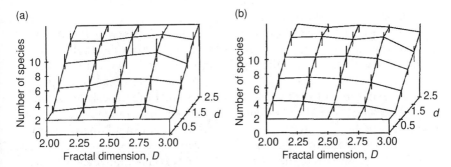

Figure 5.17 The effect of the spatial variation (d) and the fractal dimension (D) of the environment on the average and the range of the number of persistent species within the habitat, for 10 replicate runs in each case. The response curves specify relative performances in (a) fecundity and (b) competitive efficiency (after Palmer (1992)).

modified numbers of seeds n_{im} is continued until $\sum_{i=1}^{s} n_{im} = k$. α_{ij} measures the competitive effect of species j on species i. This procedure is a stochastic, discrete-event version of the Lotka–Volterra competition process (cf. also Weiner and Conte, 1981).

The species respond to the environment according to Gaussian-shaped response curves, which are equidistantly spaced out along the corresponding environmental (niche) axis. We also assume for simplicity that the response curves are identical in shape (that is, in 'width') (Figure 5.16). The response to a deviation of the environmental conditions from the optimum is a reduction in dynamical performance, in the form of a decrease either in fecundity (b_i) or in interspecific competitive efficiency ($\alpha_{ji}, j \neq i$).

The area of environmental resolution is defined as the radius r_i of the Moore neighbourhood around the home sites of species i individuals, within which they feel the local average of E. This is the most important distinctive feature of the model as compared to that of Palmer (1992), who did not consider environmental resolution, assuming that it is only the home site environment to which an individual is exposed. This explains why he arrived at the somewhat counter-intuitive conclusion that the effect of the spatial pattern of E (characterized by D) on the number of species persisting in the grid appears to be very weak, compared to the trivial effect of environmental variation (d) (Figure 5.17).

If the resolution area r_i of the species can be different, the role of the environmental pattern in determining the stationary diversity of the simulated community is much more interesting. The difference is best understood by comparing three simulated communities, differing only in the resolution area of the species: in one simulation, individuals feel only the environment of their home sites (this is Palmer's case: $r = 0$), whereas in the other two, they average E over the Moore neighbourhoods of the radius of one and two sites around the home sites ($r = 1, r = 2$), respectively (Figure 5.18). For small values of D, that is, for relatively smooth, spatially autocorrelated patterns of the environment, the difference in ultimate diversity does not appear to be significant between the communities of small and larger resolution areas. As D exceeds some threshold, the diversity of the stationary community falls rapidly with increasing r, however. The threshold corresponds to the spatial scale of environmental heterogeneity at which the environmental grain size (roughly, the average clump size of correlated sites) gets below the resolution area of the individuals. Since the sites are poorly correlated in E at large fractal dimensions, the resolution area covers a practically random sample of nine (or 25) sites, the average of which will be close to the grid average of E. The larger the resolution area, the smaller the variation in the locally averaged environments of different individuals. In other words, species with their optimum performance closest to the grid average of E have a competitive advantage and they take over, excluding other species, provided the environment is sufficiently fine-grained. Environments that are spatially heterogeneous on the scale of a single site are in fact homogeneous for species of large resolution area. This conclusion remains essentially the same, if we allow for the effects of competition other than seedling mortality (e.g., fecundity reduction).

Besides the population dynamical importance of the relative spatial scale of resolution area compared to environmental grain size, it has an interesting evolutionary ecological aspect which is worth mentioning here, being closely related to our focal topic. Suppose the resolution area is subject to heritable variation, so that plants can be selected for this property. This is not unreasonable to assume, given that environmental resolution is more or less strictly correlated with the size of the plant, which is itself subject to adaptive variation. What would be the best strategy, in terms of resolution

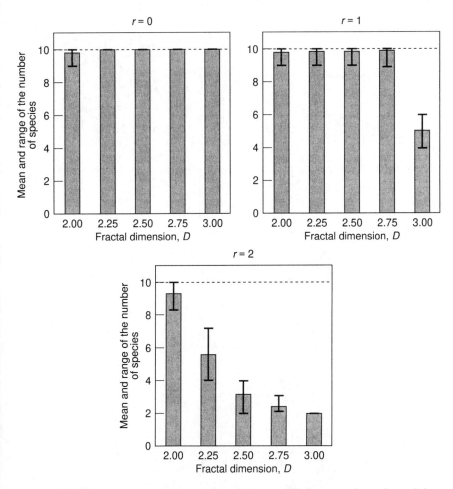

Figure 5.18 The average and the range of the equilibrium number of surviving species as a function of the fractal dimension (D) of the environment, for different neighbourhood radii (r). $d = 2.5$ in each case.

area (that is, body size) in a spatially heterogeneous environment, for a plant species to adopt? Clearly the closer the point of optimum performance of the species to the regional average of the relevant environmental variable, the more profitable it is to grow large, provided the spatial variation of the environment is sufficiently large and fine-grained. The physiological mechanism of averaging the environment over the resolution area ensures that the species can capture and exploit all the mosaic patches, including locally unfavourable ones, of the habitat. If the species performs better far

194   Site-based neighbourhood models

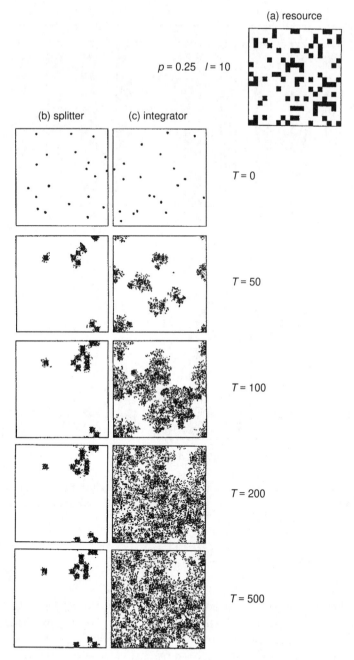

Figure 5.19 Patterns of the splitter and the integrator species at different generations of the same simulation run. Parameters: $p = 0.25$; $l = 10$.

from the regional average of the decisive environmental factor, developing a large resolution area is the worst option to chose for the plant, since it cannot even make use of the sites offering an optimal environment. It is much better to be small and to be able to exploit the scattered patches of favourable conditions. If the environmental mosaic varies in time as well, the small size of the plant should be complemented by fast reproduction and long-range dispersal, in order to be able to colonize the ephemeral habitat patches and complete their reproductive cycle before the habitat disappears. In all, part of the explanation for the wide variation in plant size on the one hand, and in reproductive and dispersal strategy on the other, may be expected to be obtained from considering the adaptation of plant species to the spatiotemporal pattern of their environment.

5.3.7 The effect of clonal integration on plant competition for mosaic habitat space

Modelling the growth geometry and the ecological interactions of clonal plants within an IPS framework is quite a new development in plant population dynamics (e.g., Ford, 1987; Oborny, 1994a, b; Cowie et al., 1995; Herben and Hara, 1995; Winkler and Schmid, 1995). (Clonal plants consist of some (often many) genetically identical, vegetatively produced units called ramets, each of which possesses everything for independent existence as a separate individual: they have a complete and functional root and shoot system. The new ramets split from the mother plant soon after having developed in certain clonal species, but in many others they remain physiologically connected through the organ of vegetative propagation (stolon, rhizome, etc.), so that the ramets of the same genetical individual (genet) can exchange nutrients and assimilates.) Since clonal growth is a process tightly constrained by steric effects (Bell, 1984; Bülow-Olsen et al., 1984; Sutherland and Stillman, 1988; Callaghan et al., 1990; Kenkel, 1995), the physiological integration of constructional units (ramets) can be expected to have an impact on the persistence and the competitive interactions of clonal plants in heterogeneous environments. Whether the effect of integration is beneficial or adverse in terms of survival and reproduction is not a trivial question – the answer must depend on the spatiotemporal pattern of the environment.

B. Oborny, Á. Kun and Sz. Bokros (unpublished results) study the role that physiological integration plays in the competition between two simulated clonal species: one called the 'splitter' (S, which eliminates the connection between the parent and the daughter ramet after the daughter ramet is established) and another called the 'integrator' (I, which maintains the connection until one of the ramets dies), in different random mosaic habitats. The model is an IPS, the arena of which is a 200×200 square grid of sites with a toroidal topology. The limiting resource is distributed in a mosaic pattern characterized by two parameters: the size of the mosaic patches (l) and the average density of the resource (p). The resource distribution is static throughout the simulation, and is determined as follows. The grid is subdivided to an orthogonal mosaic of $l \times l$ square patches, and each mosaic patch is qualified as 'favourable' with a probability p. Favourable patches are rich, unfavourable ones are poor in the limiting resource. Thus the spatial pattern of the resource is random at a scale bigger than the site scale, if $l > 1$.

The growth of the clones is initiated by sowing 40 propagules in the grid at random, 20 seeds for each species. The genets propagate by each ramet trying to put daughter ramets into the empty sites within its Neumann neighbourhood (four neighbouring sites). If there is only one candidate to colonize an empty site, it can do it unconditionally, but if more mother ramets try to send offspring into the same site, one of the daughter ramets succeeds, the others die. The chance of success depends on the resources available to the mother ramet, which differs for the 'splitter' and the 'integrator' species. An S mother on an unfavourable site has a 'vote' of weight 0.25 to delegate a daughter ramet into a neighbouring empty site; if the mother site is favourable, this weight is 0.75. I genets distribute the available resource evenly among the ramets belonging to them, so that the favourability (r) of each site below a genet consisting of n ramets, out of which f are positioned in favourable sites, is $r = f/n$. Of course, $r = 0$ if all the ramets of a genet are on unfavourable sites, $r = 1$ if all of them are on favourable ones, and $0 < r < 1$ otherwise. The vote weight of an 'integrator' ramet is $b = 0.25 + r/2$. If two or more ramets attempt to occupy the same site, one of the candidates is chosen as the winner; the chance for each competing ramet to win is proportional to its vote weight.

Each propagation update of the IPS is succeeded by a survival update, which follows the same rules with appropriate modifications: S ramets on unfavourable sites survive with a probability 0.25; those on favourable sites have a 0.75 chance to be part of the next generation. I ramets survive with a probability $s = 0.25 + r/2$. A typical run of the simulation model is illustrated in Figure 5.19.

Although the two species are identical in every respect, except for intraclonal resource redistribution, the outcome of their competition can be different, depending on the initial distribution of the propagules and the parameters of the resource pattern, l and p (Figure 5.20). When the resource is abundant (p is large), the 'splitter' species wins regardless of the grain size of the resource mosaic (l). With p decreasing, the 'integrator' species gets better, but its performance also depends on l and the initial pattern of resource capture. Of course, if p is very small, the outcome critically depends on the stochasticity of the initial propagule distribution: whichever species hits the good patches at the beginning, survives.

These results can be explained by considering how the two species behave in resource-rich and resource-poor sites. S genets are always competitively superior on a favourable patch, because they do not waste resources

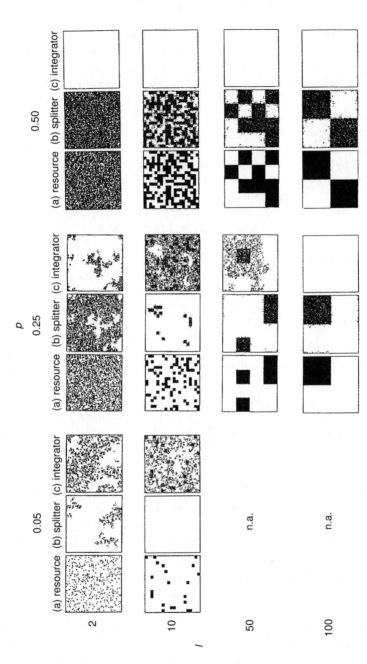

Figure 5.20 Simulation results at the 500th generation, with different (p, l) parameter combinations.

by maintaining ramets allocated to unfavourable sites outside the resource-rich patch. The smaller the patch, the faster the 'altruist' I genets are displaced. The 'splitter' species is worse in exploring resource patches that it did not colonize at the beginning of the process, however, since the dispersing ramets (those positioned outside the resource patch) die soon, and they have a small chance to reproduce. In other words, S is a bad disperser: it is bound to the patch it occupies, with a small probability of reaching another one. The 'integrator' species is better in this respect: once it gets established on a good patch, it sends out dispersers to the resource-poor area, supporting them at the expense of resources taken up by ramets in the resource-rich patch. On the other hand, the I species might have difficulties in persisting on a very small patch: with too large a fraction of its ramets falling outside the resource-rich region, the effective growth rate of the whole genet may deteriorate.

If the patches are far from each other, S is constrained onto those patches it captured *ab initio*, whereas its competitor can colonize the remaining; this regionally coexistent state can be maintained for a very long time. Such a spatially constrained steady state is not robust, however: different initial conditions result in different outcomes.

5.3.8 Percolation models of spreading populations, epidemics and forest fires

The spatial spread of populations in a patchwork habitat, disease epidemics on a population of sessile hosts, or fire within a dense forest, can be conveniently modelled within the IPS framework (Durrett, 1988), assuming that

1. vacant sites (empty habitat patches, susceptible host individuals, trees) are invaded (infected, inflamed) by neighbouring populated (infected, burning) sites with a probability p from each donor site; and
2. populated (infected, burning) sites get vacant (empty, immune or burned) in each time unit with a probability q, independently of neighbourhood configuration.

The continuous time population invasion process is in many respects similar to a Levins type metapopulation model: the sites are habitat patches, which can be in one of two states: empty (0) or occupied (1), and the local state transition probabilities p and q can be interpreted as colonization and extinction parameters. The invasion process (called the **contact process** in percolation literature) can be represented by a two-state IPS with asynchronous updating.

The simplest models of infectious disease epidemics and forest fires can be implemented as three-state IPS, the local states representing susceptible (0), infected (1) and immune (2) (or living – 0, burning – 1, burned – 2) individuals. p corresponds to the $0 \to 1$, q to the $1 \to 2$ state transition; state

2 is a sink: a site in state 2 cannot be reinfected (or re-ignited). The updating rule is synchronous, that is, time is discrete.

An interesting question from the theoretical, and possibly also from the practical viewpoint in relation to any of the above models is, what values of p and q permit the population, the disease or the fire to propagate indefinitely in space, if the process is initiated with a single site, or a small clump of sites, in state 1? Intuition says that too small values for p and/or too large ones for q will cause the process (invasion, epidemics, forest fire) to die out, but which part of the p–q phase plane allows for propagation is a quantitative problem that is not easy to solve.

The methods for answering such questions come from **percolation theory** (Kesten, 1982; Grassberger, 1983; Durrett, 1985; Liggett, 1985; Stauffer, 1985; Tautu, 1986), which deals with 'phase transitions' in discrete, stochastic spatial processes in terms of critical local transition probabilities $\{p, q\}_{crit}$. Rigorous proofs for the existence of critical values are relatively easy to provide, at least for simple processes, but finding the critical values themselves at which the process goes through a phase transition and specifying the geometry of the process are matters of lengthy stochastic simulations with a numerical implementation of the IPS in most cases. Percolation problems are critically dependent on the spatial dimensionality of the actual process, as it might be correctly guessed by intuition.

Forest fire models with more than three local states and more sophisticated updating rules have been published, e.g., by McKay and Jan (1984) and von Niessen and Blumen (1986). Notice also that the metapopulation IPS of Dytham (1994) (Section 5.3.2) and the clonal plant model of B. Oborny, Á. Kun and Sz. Bokros (unpublished results) (Section 5.3.7) can be both rephrased as percolation problems. With the recent rate of development in computer technology, a large number of more realistic population dynamical models of this sort can be expected to appear.

5.4 SUMMARY

Cellular automata and interacting particle systems are a relatively new, but also exceptionally successful type of neighbourhood – that is, spatial object-interaction – models. They are used to explain spatiotemporal phenomena in practically every branch of science, including biology in general, and population biology in particular.

CA and IPS models are ideally suited for representing the internal dynamics of complex entities composed of many spatially interacting objects; plant and animal populations and communities are obviously such entities. In ecological applications, the interacting unit is always a section of topographical space called a 'site', which is qualified by its biotic state, e.g., by the species or the species list of the organisms occupying it. Albeit the sites themselves must be simple both in the range of their possible states and in the rules of their local interactions, the collective dynamics of the whole grid of sites can be amazingly complicated. Emergent population- or community-level properties, such as static or dynamic mesoscale patterns (e.g., standing, advancing or rotating waves) or self-similar fractal structures, can be produced by very simple site-level rules, suggesting that it is not always necessary to search for subtle population-level control mechanisms or intricate site-level interactions behind seemingly complicated spatiotemporal biotic patterns.

With only a few specific exceptions, the only method to study the behaviour of a deterministic CA is numerical simulation. This applies to stochastic IPSs as well, but under certain assumptions, IPS dynamics can be simplified to a so-called mean-field approximation, which omits the local aspects of the interaction of sites, and degenerates the original object-interaction model to a mass-interaction system, taking the form of a set of difference or differential equations. Another possible type of approximation is the use of a configuration-field system, provided that the structure and the rules of the IPS model allow for the exact calculation of potential neighbourhood distributions under certain simplifying assumptions. Comparing the results of an IPS and the corresponding mean-field or configuration-field approximation, one can gain direct information about the role of spatial constraints on the given population dynamical system.

Owing to their simple formal structure and ease of computer implementation, CA and IPSs are often used as numerically convenient substitutes of reaction–diffusion systems. The two approaches differ in essential features, however, namely that CA and IPSs are discrete in space, in time and in object state, while reaction–diffusion models are continuous in all these respects. The discrete models yield qualitatively different conclusions from those of the corresponding reaction–diffusion systems in certain cases, due to the postulated particulate nature of their objects, which is an assumption obviously close to biological reality for most kinds of organism.

Besides such 'metamodels' of reaction–diffusion systems, most population dynamical applications of IPS are modified patch-abundance or patch-occupancy models, with the actual interpretation of a grid cell ranging from a small site supporting a single individual or a ramet of a clonal plant, up to the spatial scale of a landscape. Patch-abundance imitating IPSs are simplified in the representation of within-patch dynamics, but they are also extended in the sense that interpatch migration is spatially constrained (to neighbouring patches, for example). Patch-occupancy IPSs are discrete-event metapopulation systems that are made spatially explicit in the same sense; a Levins type metapopulation model is a mean-field approximation to such an IPS.

IPS models have been adapted to a wide range of population dynamical problems, including – among many others – the dynamical consequences of spatial and temporal refuge effects in competitive communities, predation in patchy habitats, competitive interactions of plants in fractal environments, and the adaptive value of clonal integration in a mosaic environment. Percolation models are a specific type of IPS, in which criteria for the spatial spread of invasive plant or animal populations, disease epidemics and forest fires can be assessed, starting from a small clump of occupied (infected, burning) sites.

6 Individual-based neighbourhood models

6.1 INTRODUCTION

Cellular automata (CA) and interacting particle systems (IPSs) define a regular (square or hexagonal) lattice of sites as the arena of interaction. The spatial discretization of Euclidean space into a regular lattice allows for all possible neighbourhoods to be identical in size and geometry, which makes the system easy to implement and manipulate on a computer. Such a spatially regular discretization is acceptable as a convenient simplification of the real topology of site-to-site interactions, as long as the model is aimed at explaining the dynamics of a system in qualitative terms like persistence or coexistence. Quantitative predictions concerning real field or laboratory situations are more difficult to obtain with CA and IPS models, however, partly because spatial discretization itself can distort the representation of biotic interactions. Even if a site-based model is correct in every other respect, the abundance trajectories it predicts might be biased from those observed on the field. This may be part of the reason why CA and IPS models are used extensively to address theoretical problems, but they are less frequently applied to imitate actual field situations.

The spatial pattern of the local biotic environment is an important determinant of the prospect for survival and reproduction of any organism. This is most obviously the case for populations and communities of sessile species like terrestrial plants. Given that the size- and distance-dependence of pairwise competitive effects can be strong in plant competition (e.g., a small difference in the distance or the spatial orientation of a competitor can have a significant effect on the vital parameters of an individual), the fate of a plant might sensitively depend on the actual neighbourhood configuration around it. With the spatial coordinates of the individuals continuously scaled and fixed (which is an assumption close to most actual field situations in vegetation studies), each individual has a unique and relatively static neighbourhood. Therefore different individuals might feel very different competitive pressures, and these differences might be frozen for a long time – possibly for generations. Spatial models designed to represent the dynamics of a sessile population or community should retain and utilize the undistorted topographical information of individual positions. Compared to site-based models, this can also mean a step further down in spatial resolution: individual-based population models are by definition small-scale in space. There are two broad types of model belonging to this class: **tessellation models** and **distance models**, the subjects of this chapter.

6.2 TESSELLATION MODELS

The word **tessellation** denotes an abstract geometric concept, which has invaded virtually all scientific disciplines in the past three decades, including many branches of physics, chemistry, biology, geography, sociology and economics. A brief definition and characterization of the concept will suffice for an understanding of the models discussed below, but the interested reader is recommended to consult Upton and Fingleton (1985) and the monograph of Okabe et al. (1992) for exhaustive treatments of the theory and some applications of tessellations.

A tessellation is a subdivision of topographical space S, generated by a set $P = \{p_1, p_2, \ldots, p_n\}$ of distinct points in S. The simplest possible tessellation assigns all those points of S to the point p_i which are closer to p_i than to any other p_j, $j \neq i$. This subdivision is called the **Dirichelet tessellation** or the **Voronoi diagram** of the generator set P in S. How the Voronoi diagram of a certain point set can be constructed in two dimensions is easy to see, considering that the points equally distant from any two distinct generator points, p_i and p_j, lie on the perpendicular bisector of the line joining p_i and p_j (Figure 6.1). The area assigned to p_i is the smallest convex polygon containing p_i, whose sides are perpendicular bisectors of the lines joining p_i

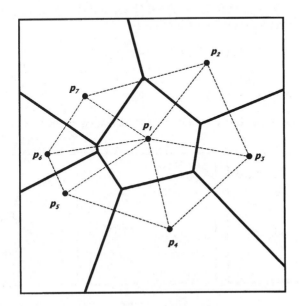

Figure 6.1 Dirichelet tessellation generated by the point set $(p_1, \ldots, p_7)$ in two dimensions.

to the other points of P (Mead, 1971). This polygon is called the **Voronoi** or **Thiessen polygon** of p_i.

The Dirichelet tessellation represents an **exhaustive** and **exclusive** subdivision of space S to such polygons; it is exhaustive in the sense that all points of S are assigned to at least one generator point and it is exclusive, because the interior points of the polygons belong to exactly one generator point: the Voronoi polygons overlap only along the common boundary lines (Okabe *et al.*, 1992). The tessellation concept and the Voronoi algorithm can be naturally extended to spaces of any number of dimensions, but most ecological applications are two dimensional (three-dimensional applications are common in crystallography and astrophysics, for example).

A tessellation is a static structure as it stands, but – based on the definition – it is easy to imagine dynamical systems that produce Dirichelet tessellations of the space. The simplest possible spatial process resulting in a Voronoi diagram is called the **simultaneous isotropic growth** (SIG) process. Consider a set of points on the plane S, each of which starts expanding (growing) at the same time t_0 with the same speed g in all directions. The growth process stops wherever two or more such 'inflating' objects (e.g., growing plant individuals) get into contact. These assumptions guarantee that each object ultimately fills the Voronoi polygon associated with the origin of its growth (that is, the corresponding generator point), so that the plane is Dirichelet tessellated in the end (Figure 6.2). With the above assumptions, the necessary and sufficient condition for a point **x** to lie on the borderline between the Voronoi polygons of individuals i and j, located at $\mathbf{x}_i$ and $\mathbf{x}_j$, respectively, is that i and j should be among the individuals who reach **x the soonest** and **simultaneously**, that is,

$$T(\mathbf{x}, \mathbf{x}_i) = T(\mathbf{x}, \mathbf{x}_j) \leq T(\mathbf{x}, \mathbf{x}_k), \quad i \neq j; \ \forall k \neq i, j, \tag{6.1}$$

where $T(\mathbf{x}, \mathbf{x}_i) = \dfrac{d(\mathbf{x}, \mathbf{x}_i)}{g}$ is the time to reach **x** from $\mathbf{x}_i$, with a speed of growth g; $d(\mathbf{x}, \mathbf{x}_i)$ is the Euclidean distance between **x** and $\mathbf{x}_i$. Within the

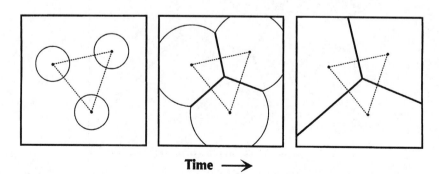

Time ⟶

Figure 6.2 Phases of the SIG process with three generator points.

borderline of i and j, but outside the vertices (the 'corner' points of contact for more than two polygons), the inequality is strict; at the vertices the equality holds for those polygons that are in contact, and the strict inequality holds for all other points of S. A numerical realization of the SIG process and astrophysical applications are given in Zaninetti (1989).

The Dirichelet tessellation is an attractive paradigm of individual-based population dynamical modelling in at least two respects. One is that it assigns a section of space to every individual, the extension and the shape of which is very sensitively dependent on local population density and the geometry of the neighbourhood. If resources are homogeneously distributed in space, then the area of the Voronoi polygon around an individual is a measure of the quantity of resources closest to, therefore possibly the best available for that individual (Brown, 1965; Firbank and Watkinson, 1987). The other advantage is that the individuals whose Voronoi polygons have a common edge with that of the focal individual can be naturally considered as neighbours. There is no need to apply either arbitrary prior assumptions on the spatial arrangement of the interacting objects, such as a square or a hexagonal grid, or parameters that might be hard to measure or to interpret, in order to determine which individuals do and which others do not belong to the neighbourhood of a given individual. Moreover, the growth model defined above is directly capable of modelling plant foliage or root competition for space, under the assumptions that the foliages (or the root systems) grow isotropically and uniformly, and they cannot be penetrated by each other.

The limitations of tessellation models in ecology are related to the facts that (i) their use is more problematic in modelling the population dynamics of mobile species, although they can be naturally applied to organisms that are sessile in the broad sense (including plants, benthic and territorial animals), and that (ii) individuals outside the immediate neighbourhood are excluded from among those directly affecting the survival and regeneration prospects of the focal individual (Pacala and Silander, 1985) – at least in the most appealing, simple models. Cannell *et al.* (1984) show that (ii) is not a serious limitation in the case of the interaction of two conifers, *Picea sithensis* and *Pinus contorta*: arranged in a hexagonal grid, the size of the six immediate neighbours predicted the intensity of the competitive interaction on the focal tree, and the prediction was not any better if more distant individuals were considered as well. Hara (1985) came to a similar conclusion in a different study, finding that the number of neighbours to consider for the best prediction of performance is almost always very close to six, which is the expected number of neighbours in the Dirichelet tessellation of a random point pattern (Upton and Fingleton, 1985; Okabe *et al.*, 1992). These, and a number of other studies (see Kenkel (1990) for a review) suggest that the state of the neighbourhood as defined by the Voronoi diagram can be a good predictor of the biotic effects on the focal plant, at least in even-aged stands.

6.2.1 Predicting plant performance from tessellation parameters: the Voronoi assignment model

Tessellations are most widely applied in plant ecology as single-species models of pattern-dependent competitive interaction. The overwhelming majority of such plant competition models are designed to predict the outcome of, and most of them are actually tested against, certain field processes or laboratory experiments. They are different from the usual individual-based population dynamical systems in a very important respect, however: the **time dimension** is implicit, meaning that the abundance trajectories are in fact not followed through generations. The aim is to forecast the end state of a short-term process from its initial state, regardless of what series of intermediate states the process passes through. Such **temporally implicit** tessellation models are called Voronoi assignment models (VAM) by Okabe *et al.* (1992).

To be more specific: the VAM approach is applied to predict plant performance (in terms of survival, fecundity, rate of biomass increase or other vital parameters) at a certain period of the growing season from the actual or an earlier pattern of the same stand. The local biotic environment of each individual is characterized by the parameters of the Voronoi polygon (called the **plant polygon**) around it. The most widely used parameters are the **area**, the **shape** of the plant polygon of the focal individual, and also the **number** and the **proximity** of its neighbours. The method of relating plant performance to the Voronoi parameters of the pattern is statistical in all cases: the distributions of tessellation parameters are correlated to those of plant performance. The basics of the statistical methodology of VAM models have been worked out by Mead (1967, 1971).

In his pioneering paper, Mead (1966) studied carrot monocultures from sowing to harvest, and found that three parameters of the Voronoi polygons:

1. area
2. excentricity (an anisodiametry measure of the plant polygon) and
3. abcentricity (the distance of the plant from the centroid of its polygon)

together account for more than 20% of the variation in plant yield at harvest. Not surprisingly, plants located close to the centroids of large, isodiametric plant polygons can be expected to produce the largest biomass.

Quite a large number of field and laboratory studies on plant monocultures (and, in a few cases, on two-species systems) along the lines initiated by Mead (1966) have been published in the past thirty years, applying the VAM methodology to many different species (e.g., Fischer and Miles, 1973; Mack and Harper, 1977; Liddle *et al.*, 1982; Watkinson *et al.*, 1983; Mithen *et al.*, 1984; Hutchings and Waite, 1985; Matlack and Harper, 1986; Firbank and Watkinson, 1987; Kenkel, 1988; Kenkel *et al.*, 1989; Owens and Norton, 1989; Aguilera and Lauenroth, 1993). The basic situation is similar in all these studies: starting from a dense, more or less even-aged, natural or artificial stand of a herb or a tree species, plant performance parameters – measured in different phases of the self-thinning process – are related statistically to certain parameters of a Voronoi tessellation of the habitat. The tessellations can be generated by the pattern of seedlings (e.g., Watkinson et al., 1983; Mithen et al., 1984; Hutchings and Waite, 1985), adult plants (e.g., Mead, 1966; Aguilera and Lauenroth, 1993) or both (e.g., Liddle et al., 1982; Matlack and Harper, 1986). The generating pattern is the reference state to which plant performance is correlated. In certain cases, the reference state can be the pattern at the time of performance measurement, that is, plant performance is related to the actual pattern (e.g., Mead, 1966). From a dynamical viewpoint, this is a reasonable method only if the positions of individuals that died out from the pattern can be determined; thus their performance can also be considered in the statistics.

The amount of variation in plant performance that can be explained by the Voronoi parameters seems to range from about 20% (Liddle et al., 1982) to 60% (Waller, 1981) in different cases, partly depending on how the 'all other things equal' criterion could be satisfied in a particular field or laboratory setting. Most of the difficulties in this respect stem in asynchronous seedling emergence, which is partly due to the genetic variability of the seeds sown. Watkinson et al. (1983) demonstrated that the temporal order of seedling emergence appears to explain most of the variation in performance, due to the **temporal pre-emption** effect in resource capture. Plants emerging sooner can extend their roots and foliage to occupy a disproportionately large area, and thus they can use a disproportionate part of the resources compared to later emerging individuals. The temporal pre-emption effect is strong, and it does not have much to do with spatial constraints. (As will be shown later, the Voronoi growth model can be generalized to incorporate the temporal pre-emption effect, cf. Section 6.2.6.) In more or less synchronously germinating and genetically uniform populations, the spatial pattern of the seedlings explains much of the variation in performance. In fact, plants with a large Voronoi polygon around them enjoy a **spatial pre-emption** effect, escaping from the competitive effect of neighbours for a longer time than other individuals with a smaller polygon. The general conclusion that plant performance is more or less correlated (at least in the competitive phase of the dynamics) with plant polygon area and the number of neighbours, seems to be quite robust throughout the literature. Polygon shape, abcentricity and other Voronoi parameters show a much less consistent effect between different studies.

6.2.2 An interpretation of the self-thinning rule on the individual level

An interesting link between the so-called '–3/2 power law' of self-thinning (Yoda et al., 1963; Harper, 1977; Begon et al., 1986) and the tessellation model of an experimental stand of *Lapsana communis* has been revealed by Mithen *et al.* (1984). The −3/2 power law is one of the few, relatively robust relationships of plant ecology, stating that in a densely sown, even-aged monoculture the number of individuals decreases, and the biomass per plant increases with time in such a way that the points of the log (mean plant biomass)–log (density) plot, recorded at different times through the thinning process, approach a straight line of −3/2 slope (Figure 6.3). This amounts to the algebraic form of the law:

$$w = C \cdot d^{-\frac{3}{2}}, \tag{6.2}$$

where w is mean plant biomass, d is density (that is, plants per unit area) and C is a species-specific constant. This relation seems to be applicable to vascular plants almost independently of their actual taxonomic identity, as it has been found for many species with different constants from small herbs

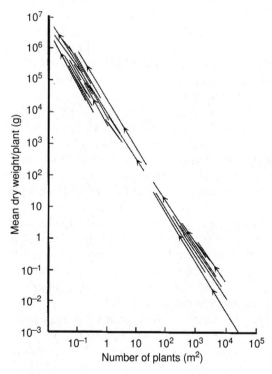

Figure 6.3 Thirty-one experimental and field cases of the self-thinning process, including data for various herbs and trees. Note the striking adherence of all self-thinning trajectories to the same line of slope −3/2, which demonstrates that the allometric relation of plant territory and plant size applies to many plants, regardless of their species identity and growth form (after Begon *et al.* (1986)).

to trees since Yoda et al. (1963) observed it for the first time. The interesting result of Mithen et al. (1984) is that in a saturated, self-thinning *Lapsana* monoculture the regression line of the log–log plot of plant dry weight (as a measure of performance) against plant polygon area has a slope of 3/2 (Figure 6.4). Translating the graph of Figure 6.4 to an algebraic expression, one gets

$$w = C \cdot A^{\frac{3}{2}}, \tag{6.3}$$

where W is plant weight and A is plant polygon area. Note that the reciprocal of plant polygon area (which is of dimension area/plant) is a local density measure (of dimension plant/area): $d = A^{-1}$, so that (6.2) and (6.3) are in fact identical.

Just as in the case of the self-thinning law, it is possible to interpret this result on the individual level, even if the relevant physiological mechanisms are not known in detail. Given the astonishingly wide applicability of the self-thinning law, one might expect the explanation to be connected to the most universal morphological–architectural and physiological properties of vascular plants. Specifically, it seems plausible to assume, for example, that in a crowded even-aged stand the amount of resources available to a plant individual is contained within the section of space above and/or below the plant polygon. The volume of this section of space is proportional to $A^{\frac{3}{2}}$, the coefficient of proportionality depending on many species-specific properties like the shape and the structure of the root system or the foliage, the physiological efficiency of resource capture and utilization, etc. If we further

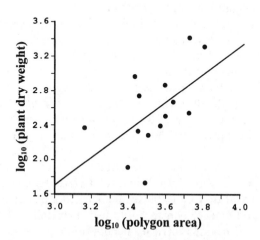

Figure 6.4 The relation of plant dry weight and plant polygon area in a self-thinning experiment with *Lapsana communis*. The equation of the regression line is $y = \frac{3}{2}x - 2.8$; $r^2 = 0.30$; the regression relation is significant at $p < 0.05$ (after Mithen et al. (1984)).

assume that the size of a plant is approximately proportional to the amount of the limiting resource (light, nutrients or water) available to it, we end up with (6.3) as the expected outcome of the self-thinning experiment. The size-to-resources proportionality assumption can be true only within certain limits, of course, since there is obviously a minimum viable size of individuals, and a genetically and physiologically determined maximum size that can be reached under optimal environmental conditions. With these constraints, the same simple reasoning can be applied to the self-thinning law (6.2) as well, which differs from (6.3) only in that the dependent variable is mean plant weight, w, instead of individual plant weight, W. It may be that only this specific case of *Lapsana* conforms to the $-3/2$ exponent rule on the individual level as well, or it may be a more general law; this matter requires further studies on other plant species before a conclusion can be reached.

6.2.3 Tessellation models of territory establishment

Although the tessellation approach is not conveniently applicable to populations of mobile organisms in general, there are a few important exceptions (see, for example, Hamilton (1971) or Cannings and Cruz Orive (1975)). The most obvious ecological situation in which the Voronoi diagram is a good approximation to the actual pattern of animal populations is that of competition for territories. If each individual defends a part of the habitat as its own territory, excluding all potential conspecifics from that area, and the whole area is covered with territories, the resulting population pattern can be approximated by a tessellation of the habitat with the nests or home sites being the generator points. One of the most spectacular realizations of such polygonal territories has been observed in an experiment by Barlow (1974). He put male individuals of the mouthbreeding cichlid fish *Tilapia mossambica* into an artificial outdoor pool that had a uniform sandy bottom. The bottom of the pool was divided by the fish, who excavated breeding pits by spitting the sand out from the centre of the pit towards nearby individuals, who did just the same. If the density of the fish population was large enough, the result of the conflicting nest-building efforts was a network of sand parapets. This network formed a pattern very similar to a Voronoi diagram generated by a more or less evenly spaced out point set.

The fact that the individuals were ultimately evenly distributed in the pool suggests that the fish adjusted their territory centres during the tessellation process so that they be as far from each other as possible. Tanemura and Hasegawa (1980) have constructed the dynamical model of this situation, assuming synchronous territory occupation and a subsequent spatial readjustment process. They started with a random point pattern; the random vector of spatial coordinates for the ith individual at time 0 was $x_i(0)$. Then they assumed that the fish adjusted their positions in space to decrease their abcentricity within the territory, that is, their distance from the areal centre of gravity (centroid) $\mathbf{g}_i$ of the Voronoi polygon assigned to $\mathbf{x}_i$. The algorithm of the adjustment process was very simple:

$$\mathbf{x}_i(t+1) = \mathbf{x}_i(t) + \frac{\mathbf{g}_i(t) - \mathbf{x}_i(t)}{M}, \tag{6.4}$$

where M defines how smoothly (slowly) the individual adjusts its actual position closer to the centroid of its territory – the larger the value M takes, the smoother the process. The Voronoi diagram is regenerated at every time step with the modified generator set, that is, the centres of gravity (polygon centroids) also move in time. The territory pattern of the habitat converges towards a time invariant (stable) configuration, with the distances between neighbouring generator points being almost the same everywhere (Figure 6.5).

The Voronoi adjustment process cannot work in cases of asynchronous arrival and territory establishment, if the individuals fix their positions immediately after having checked that their distance from the nearest resident neighbour is larger than a critical minimum value. Nest-building territorial birds are examples for this type of space division, because once determined by the bird or the bird pair, the position of the nest can hardly be altered. Tanemura and Hasegawa (1980) consider this case as a 2D sequential random packing problem, where the objects to be packed are circles of diameter d, which is the minimum required distance from the nearest neighbour. Assuming that the birds invest equal effort in defending their territories, the resulting pattern of territories – after all the nests are established – is a constrained Voronoi diagram, where each polygon includes a circle of diameter d (Figure 6.6). Field data of Grant (1968) and Buckley and

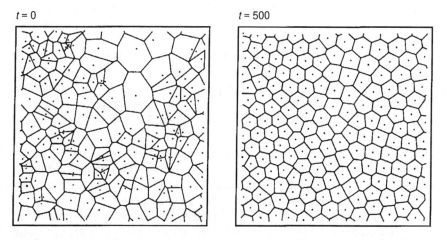

Figure 6.5 The (random) initial state ($t = 0$) and the equilibrium pattern ($t = 500$) of the territory adjustment process (after Tanemura and Hasegawa (1980)).

212 Individual-based neighbourhood models

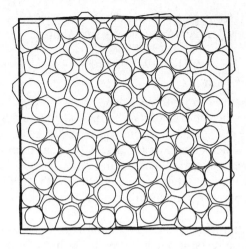

Figure 6.6 The constrained tessellation model of territory packing with a minimum diameter of the Voronoi polygons (after Tanemura and Hasegawa (1980)).

Buckley (1977) for different bird species fit this model of territory occupation quite well.

6.2.4 Towards tessellation dynamics: linking tessellations to demography

From a theoretical point of view, the principal novelty of the adjustment model and the sequential random packing model of territoriality, as compared to the Voronoi assignment models of Sections 6.2.1–3, is the explicit time dimension considered. However, the relevant time scale of both territory models matches the duration of the territory establishment process, not extending over a single season. Population dynamical models in the strict sense, that is, those which focus on the long-term abundance dynamics of the population(s), are still missing from among tessellation models, however. This ignorance on the part of population dynamics is quite surprising, given that most tessellation models use the area of the Voronoi polygon as a predictor of the individual's performance, which is actually a dynamical parameter in many cases, such as survival or fecundity. Thus the output of a tessellation model (regardless of whether it is static like assignment models, or dynamic like the territory models) offers itself as the potential input of a linked population dynamical system. The reason for the ignorance thus far may be technical: tessellation models require the excessive use of numerics and even static tessellation algorithms might be rather slow, if the number of generator points is large. This technical limitation tends to disappear with the fast improvement in computer technology, but another problem is that tessellation algorithms are very non-trivial to program.

Tessellation models   213

A real dynamical tessellation model for the self-thinning of an even-aged tree monoculture was developed by Galitsky (1990). The key variables of his model are individual plant biomasses in the physiologically active and the inactive parts, which increases in time as a function of the actual plant polygon area: a larger plant polygon allows for faster growth. Any plant whose biomass falls below a certain threshold (a fixed proportion of the free-growing plant's biomass of the same age) dies immediately, and its 'territory' is shared among its neighbours as if it had never been present. The tessellation is recalculated after every death event. With appropriate growth functions and threshold values, this model could be tested against some of the many field and laboratory studies on self-thinning, for both the size distribution and the spatial pattern of the surviving plants, but this is still to be done. The model was not tested for whether it is capable of producing the $-3/2$ power rule – this is also an interesting question to study with the same, or with a similar, tessellation model.

Even this model, although fully dynamic, considers death as the only demographic process of the population. Fecundity and dispersal, which could establish the dynamical connection to the next generation, are not pondered at all. To my knowledge at least, there are no tessellation models including both birth and death processes in plant or territorial animal populations as yet, albeit this extension of the tessellation approach seems to be straightforward and inevitable in the future. Dynamical tessellations with this modification would become a new type of individual-based models on the small spatiotemporal scale. Beyond doubt, such models could be interesting for both the theoretician and the field ecologist.

6.2.5 Towards multispecies tessellation dynamics: weighted tessellations

Having a single-species tessellation approach at hand, it is a natural wish to extend it to multispecies systems as well. This is a relatively straightforward extension in principle: if the generator points are assigned species identity, the tessellation algorithm can be generalized in many different ways to represent interspecific, that is, asymmetric interactions besides intraspecific (symmetric) ones. The methodological framework of generalized, asymmetric tessellations has been worked out (see Okabe *et al.* (1992) for a detailed review) and it is also applied in other disciplines (they are, for example, standard tools in geographic information systems (GIS); see Okabe *et al.* (1994)).

For possible plant population dynamical applications, perhaps the most straightforward generalizations of the Voronoi diagram can be derived by introducing **weighted distances**, and assuming that plants have access to resources in those parts of the habitat that are closest to them in terms of the weighted distance. Suppose, for the simplest example, that g, the speed of reaching a certain point $\mathbf{x}$ of the habitat from the generator point i at $\mathbf{x}_i$

214 Individual-based neighbourhood models

depends on the species identity s_i of i. We assign all those points of the habitat to the 'territory of resource exploitation' of the plant rooted at $\mathbf{x}_i$ which can be reached the soonest from $\mathbf{x}_i$ with that particular speed g_{s_i}. The resulting tessellation pattern (Figure 6.7) is again an exhaustive and exclusive division of the plane, but it is quite different from an ordinary Voronoi mosaic otherwise. The tiles are not polygonal, since the borders between individuals of different species (with different speeds of growth) are not straight, and the area of a tile depends on the species in the generator point and those in its neighbourhood. The pattern of plant 'territories' in an even-aged plant community might be such, if the species of the community differ in the horizontal growth rates of individuals. Formally, the topographical distance $d(\mathbf{x}, \mathbf{x}_i)$ of a point $\mathbf{x}$ from an individual i of species s_i located at $\mathbf{x}_i$ is weighted with the reciprocal of the speed of growth g_{s_i}, so that

$$T_{s_i}(\mathbf{x}, \mathbf{x}_i) = \frac{1}{g_{s_i}} d(\mathbf{x}, \mathbf{x}_i). \tag{6.5}$$

$T_{s_i}(\mathbf{x}, \mathbf{x}_i)$ is the time that an individual of species s_i needs to reach $\mathbf{x}$ from $\mathbf{x}_i$; this is called the **multiplicatively weighted distance** of $\mathbf{x}$ from $\mathbf{x}_i$. The borders of the tiles consist of the points equidistant from two or more individuals,

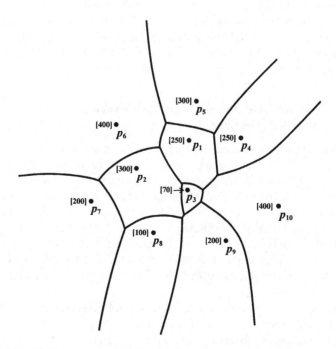

Figure 6.7 Multiplicatively weighted tessellation generated by a set of 10 points. Each square bracketed number is the multiplicative weight g_i (speed of growth) associated with the generator point p_i nearby (after Okabe *et al.* (1992)).

like in the case of the ordinary Voronoi diagram, but now in terms of the multiplicatively weighted distance (that is, growth time). The condition for point **x** of the habitat to lie on the borderline between the tiles of individuals i and j, belonging to species s_i and s_j, respectively, is

$$T_{s_i}(\mathbf{x}, \mathbf{x}_i) = T_{s_j}(\mathbf{x}, \mathbf{x}_j) \leq T_{s_k}(\mathbf{x}, \mathbf{x}_k), \quad i \neq j; \ \forall k \neq i, j. \tag{6.6}$$

If all values of g_s are equal, the resulting tessellation is the ordinary Voronoi diagram.

The multiplicatively weighted model simulates the pre-emption mechanism of fast 'territory capture' in an attractively simple way. There is another possible pre-emption mechanism, namely the temporal order of germination (temporal pre-emption, cf. Section 6.2.2), which is one of the most important factors determining the performance of individual plants, both in monocultures and in multispecies communities. In fact, most VAM studies of field and laboratory data with different species led to the conclusion that the timing of germination is the best predictor of future performance; the spatial parameters of the tessellation tend to be secondary to it (e.g., Liddle *et al.*, 1982; Watkinson *et al.*, 1983; Firbank and Watkinson, 1987): the sooner the plant emerges, the better it performs later, because it can extend its influence to a larger area and thus it might get a larger share of the resources.

The differences in emergence time can be included in another type of weighted tessellation system by dropping the synchrony assumption. Let seedlings emerge at different times, and let them start occupying territory with the same constant speed g (that is, no multiplicative weighting is considered for the moment). Let t_i be the time elapsed until the germination of individual i, relative to a reference point of time t_0 (t_0 should be conveniently chosen so that it precedes the first germination event, that is, $t_0 < t_i$). Then the time from t_0 to the event that i reaches **x** from $\mathbf{x}_i$ is

$$T_i(\mathbf{x}, \mathbf{x}_i) = \frac{d(\mathbf{x}, \mathbf{x}_i)}{g} + t_i, \tag{6.7}$$

which is the **additively weighted distance** of **x** from $\mathbf{x}_i$.

Additively weighted tessellations are also space-exhaustive and space-exclusive; the border between the tiles of points i and j is also curved if $t_i \neq t_j$, but, in addition to these features, the tiles might be fragmented as well (Figure 6.8). The latter property is attributable to the fact that we do **not** assume that growth stops when two or more growing tiles get in contact; instead, we simply say that a tile consists of those points of the plane from which the additively weighted distance (6.6) of the corresponding generator point is the smallest compared to that of all other generator points. This property can hardly be interpreted for solitary organisms, but it is more realistic as a representation of the growth of clonal plants, for which g might represent the speed of the spatial spread of ramets by vegetative spacer

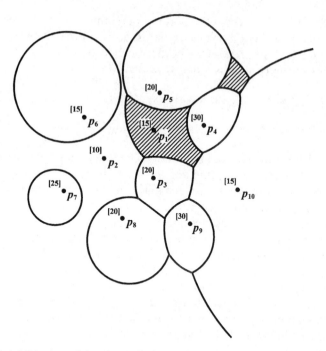

Figure 6.8 Additively weighted tessellation generated by the same set of generator points as in Figure 6.7. Square bracketed numbers are the additive weights, t_i (time lag of germination). Both hatched tiles belong to p_1 (after Okabe et al. (1992)).

organs such as stolons or rhizomes. The model is applicable only if it is also assumed that spacers do not severely constrain each other, that is, the presence of one does not stop the growth of another. There can be tiles of zero area, if the emerging plant finds itself within the tile of a neighbour that got there sooner, representing an extreme case of temporal pre-emption. With all values of t_i equal, the model yields the ordinary Voronoi diagram.

Multiplicative and additive weighting can be combined in the same model. The **compoundly weighted distance** T_{i,s_i} of an individual of species s_i, located at $\mathbf{x}_i$ and emerging with a time lag t_i after the appearance of the first seedling in the community, from point $\mathbf{x}$ of the plane is

$$T_{i,s_i} = \frac{d(\mathbf{x}, \mathbf{x}_i)}{g_{s_i}} + t_i. \tag{6.8}$$

The resulting tessellation is even more complicated than that in the additively weighted case (Figure 6.9), and it is even less straightforward to interpret as a realistic plant interaction model, because it is also possible that the 'territory' of a clone does not contain the point where the mother ramet emerged, that is, the plant has to be assumed to develop spacers without the

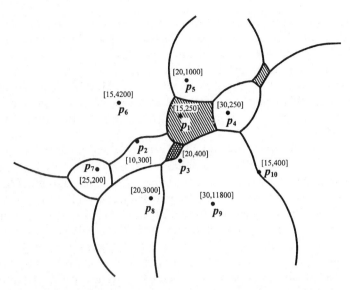

Figure 6.9 Compoundly weighted tessellation for the same set of generator points as in Figures 6.7 and 6.8. Square bracketed pairs of numbers are the additive weights and the multiplicative weights respectively, associated with the corresponding generator point. Left-hatched tiles belong to p_1; the cross-hatched tile belongs to p_3 (after Okabe *et al.* (1992)).

mother ramet being actually established. The problem could be offset by adopting the key assumption of the SIG model (cf. Section 6.2), namely that the growth of a tile ceases when and where it gets in contact with one or more other tiles. This also excludes the appearance of fragmented tiles in the additively and compoundly weighted models. The SIG algorithm defines another model, however, which is different from the additively and compoundly weighted tessellations in that plant territories do in fact interfere with each other. Note that in the multiplicatively weighted model, just like in the SIG model itself, it makes no difference in the algorithm and the resulting pattern whether we define tile borders in terms of simple weighted distances or in terms of the SIG process of interfering tiles.

In population dynamical applications, the individual differences in the weighting factors can be flexibly interpreted according to the actual problem, either as within-species phenotypic differences, or as differences between species. The weighting factors ($1/g_{s_i}$ and t_i) are in fact *i*-state variables, which can take different values for each individual, or they can be canonical with other *i*-state variables like species identity. Between-species and within-species differences can be combined quite freely. For example, there can be an interspecific difference in the timing of emergence, but it need not be uniform within the species either. If they are not, then the means of the intraspecific distributions of t_i might characterize between-species differences; the variances can be used as measures of within-species variability. The same may apply to the speeds of growth as well. Of course there are a number of other weighting methods (see Kenkel (1990) for a review of some) used only in static VAM studies so far – many of these are much more difficult, if possible at all, to interpret in terms of individual growth and interference. Their use in population dynamical models is therefore more problematic.

Even the multiplicatively and additively weighted tessellations have not yet been used in multispecies population dynamical models, albeit the extension of the tessellation approach to competitive interactions in plant (and other sessile) communities seems straightforward. Weighted Voronoi methods could be relatively easily applied in individual-based growth models, which in turn can be used to determine the small-scale, short-term demography of the interacting species in specific field cases as well. All such models must be definitely computer-orientated, because the formalism associated with tessellations is largely intractable analytically, but – as numerical implementations can be detailed (and thus realistic) enough – the model outcomes could be directly compared to field or experimental data in exchange – and this is a regrettably rare virtue among analytical models in general.

6.3 DISTANCE MODELS

The tessellation approach to individual-based population dynamics rests on the assumption that topographical space, and thus all the resources associated with it, are divided among the individuals in a space-exhaustive and space-exclusive manner; that is, any part of space is assigned to, and utilized by, one and only one individual. The exhaustive division of space is a realistic assumption if the density of individuals is large enough everywhere for each individual to be constrained by its neighbours in resource uptake and growth. For territorial animals, this can be true within quite a broad range, because the segment of space that an animal can exploit and defend might be large compared to its body size, and the territories can be also rather flexible in size, depending on the actual population density in the area. For plants, however, the exhaustivity assumption means that the actual tile of each plant should be contained within the area that the plant could utilize had it been grown alone – and this potentially occupied area is usually not much larger than the body size (foliage or root system diameter) of the plant itself. In other words the tessellation approach is nicely applicable in saturated populations and communities, but it might be very complicated to tailor a tessellation model to fit to plant competition processes in unsaturated habitats as well. The occasional appearance of very abcentric tiles can also be problematic, if abcentricity is not accounted for explicitly in a tessellation model, because there are limits to the geometric plasticity of plants:

the roughly circular (or at least roughly isodiametric) ground projected shape of the foliage or the root system of a single viable individual cannot usually be distorted to an arbitrary measure – plant territories are not very flexible in shape.

The assumption that the area of a tile belongs to only one individual is also questionable in many cases. It is quite obviously violated by many herbs, whose root systems can be so intimately entangled that it may seem hopeless to separate them by any means. Sometimes the same applies to the above-ground parts as well.

None of these basic assumptions of the tessellation approach is unavoidable in **distance models**, which are an alternative class of spatially explicit individual-based systems. The name of this class refers to a general property of the models it includes: the interaction among the individuals of a population or a community is represented by a direct function of their topographic distance. There are basically three types of distance models, differing in the specific functions which define neighbourhood relations and determine the intensity of the interaction:

1. **Fixed radius neighbourhood (FRN) models.** Each individual is in the centre of a circular disque of fixed radius; all other individuals occurring within that disque are neighbours, influencing the chance of survival and/or the fecundity of the focal individual. The effect of a neighbour might depend on its species, age, or other i-state descriptors; sometimes it is also dependent on the actual distance from the focal plant. The most elaborate examples of this kind of model were published by Pacala and Silander (1985) and Pacala (1986a, b, 1987).
2. **Zone of influence (ZOI) models.** Each individual is supposed to exploit resources within a circular zone (ZOI) around its stemming point. Resource competition is assumed to occur wherever the zones of influence of two or more individuals overlap. The intensity of the competitive effect on a plant might depend, for example, on the fraction of its ZOI overlapped. This fraction can be in many ways weighted, with the actual distance of the overlapped fraction from the stemming point, or otherwise. The ZOI of an individual can obviously be i-state dependent. Examples for ZOI models are Czárán (1985) and Czárán and Bartha (1989).
3. **Ecological field (EF) models.** The principal difference of EF models from the previous two types is that they do not require that the plant individuals be assigned arbitrary neighbourhood radii or zones of influence, outside of which no interaction is possible. Instead, each plant individual is supposed to have an ecological impact on the habitat such as shading, water or nutrient depletion, allelopathic effects, etc., or any combination of these. The environmental impact is represented by a (usually monotone) function of distance from the plant, so that the effect on the environment might be weak but still positive at large distances. The

individual impacts are superimposed to yield the joint pattern of ecological impacts, or the **ecological field** within the habitat. The local field values are then used to determine the survival and/or the reproductive output of each individual. EF models have been introduced by Wu *et al.* (1985), and later developed by Walker *et al.* (1989).

6.3.1 Fixed radius neighbourhood models

The key assumption of FRN models is that the plant individuals stemming within a certain distance from the focal plant are responsible for the biotic effects that the focal individual suffers. Moreover, the survival probability and the expected fecundity (that is, the performance) of a plant is assumed to be predictable with reasonable certainty, if the i-state configuration of its neighbourhood is specified. Starting from these postulates, Pacala and Silander (1985) and Pacala (1986a, b, 1987) have developed a series of plant population dynamical models, the range of which extends from single-species systems of annuals to multispecies models with spatially heterogeneous abiotic environments. I shall outline their single-species models for the competitive interaction of annuals below, and also briefly touch upon the multispecies extensions to this model, including both computer simulations and the corresponding analytical configuration-field (cf. Section 5.3.3) approximations. I preserved the notations of the original papers as much as possible, to make it easy for the reader to turn to the original papers for more details or derivations.

Starting from the single-species case, assume that the performance of each plant depends only on the local density around it, that is, the number of conspecifics within its FRN. This dependence is specified by the so-called **performance predictor functions**, which can be either reasonable arbitrary functions, or else they can be obtained from experimental or field data, using regression statistics. We assume further that it is only the **chance of survival** and **fecundity** that depends on local density, so that there are two density-dependent predictors to be actually specified: the **survival predictor** and the **fecundity predictor**.

If the survival predictor $z(n)$ is obtained from field data, the procedure starts with mapping the rooting points of all individuals of the sample site into computer memory. The subsequent procedures can be automated: the computer determines the number of individuals falling within the FRN of each individual for any neighbourhood radius r, and plots the proportion z of survivors among those seedlings that had exactly n seedling neighbours within their FRN. The resulting plot is the empirical survival predictor. It can be used as it is in numerical studies, but for analytical methods to be applicable, the survival predictor must be defined in closed form, that is, it is necessary to fit an analytical function to the points of the graph. The radius r of the neighbourhood is a matter of more or less arbitrary choice, but it should not be very different from the horizontal size of an individual.

Distance models 221

In the case of a fitted predictor, it is reasonable to chose the radius giving the best fit of the predictor function to the actual data. Irrespective of whether it was arbitrarily chosen or fitted, the survival predictor function expresses the negative density dependence of survival, that is, in analytical form it is a decreasing function of the local density n. A possible option for a closed formula would be, for example, $z(n) = Be^{-an}$, where B and a are constants.

The fecundity predictor $f(n)$ can also be obtained from field data by plotting the number of seeds set by each plant against the number of neighbours n within its FRN, and fitting an analytical function to the cloud of points thus produced (Figure 6.10). n might be either the number of neighbours in the seedling phase (before density-dependent mortality occurred), in which case the fecundity predictor is called a **seedling fecundity predictor** (SFP), or else it can be the number of adult neighbours (those remaining after density-dependent seedling mortality), whence it is an **adult fecundity predictor** (AFP). $f(n)$ should also be a decreasing function of n,

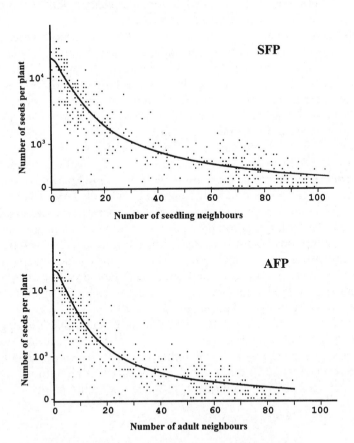

Figure 6.10 Experimentally obtained seedling and adult fecundity predictors (SFP and AFP) of *Arabidopsis thaliana* (after Pacala and Silander (1985)).

possibly like the exponential function in the example for $z(n)$ above, or a reciprocal power function such as $f(n) = M/(1 + cn^\tau)$ like the fitted curve in Figure 6.10, among others. The radii of the effective neighbourhoods need not be the same for the survival and the fecundity predictor, but for sake of simplicity I will assume they are (Pacala and Silander (1985) derive analytical results for the more general case as well).

There are two things left to be determined for a complete spatiotemporal life history specification of the species. One is the spatial locations of the seeds produced by a mother plant (dispersal pattern), the other is the probability of germination. The dispersal pattern can be again a distribution fitted to experimental data – or else it can be chosen arbitrarily, a convenient option being, for example, the 2D Gaussian distribution centred on the position of the parent plant; the dispersal parameter of the model can be the standard deviation of the Gaussian in this case. Germination probability might also depend on local crowding, so that a separate germination predictor function $g(n)$ could be produced, but for simplicity it will be assumed to be a density-independent constant g in the forthcoming.

With the performance predictors $z(n)$ and $f(n)$, the dispersal mechanism and the germination probability specified, the spatiotemporal dynamics of an annual plant population is easy to simulate by computer. Starting from any initial pattern of seedlings, the survival probability and the number of germinated seeds can be calculated for each plant, and the locations of their seedling progeny can be determined by drawing them from the dispersal distribution centred on the mother plant.

It is possible to obtain analytical results for this system if the survival predictor and the fecundity predictor are known in algebraic form, with a few restrictive assumptions made on habitat size, initial pattern and dispersal. Specifically, suppose that (i) the habitat is infinitely large, (ii) the initial pattern of seedlings is random and (iii) the distance of seed dispersal from the parent plant is large as compared to r, the radius of the neighbourhood. If these assumptions hold, the spatial pattern of the seedlings will be indistinguishable from random in all generations (see Pacala and Silander (1985) for a proof). In other words, assumptions (i)–(ii) guarantee that an appropriate configuration-field model is a good approximation to the spatiotemporal dynamics of the plant population.

The most general form of the model simply states that the expected average density S_{t+1} of seeds at time $t+1$ is a product of the average seedling density gS_t at time t (g is the probability that a seed will germinate), the average probability $Z(S_t)$ that a seedling will survive to adulthood and the average number $F(S_t)$ of viable seeds produced per adult plant:

$$S_{t+1} = gS_t \cdot Z(S_t) \cdot F(S_t). \tag{6.9}$$

To derive explicit algebraic expressions for the expected survival $Z(S_t)$ and the expected fecundity $F(S_t)$, recall that the spatial distribution of seedlings is Poisson distributed at all times, so that the probability that a neighbourhood of radius r will contain exactly n seedling neighbours is given by

$$p_n = \frac{\lambda^n}{n!} e^{-\lambda}. \tag{6.10}$$

p_n is the nth term of the Poisson distribution, the parameter $\lambda = gS_t \cdot r^2\pi$ of which is the average number of seedlings per neighbourhood. Rescaling both spatial dimensions in units of $r\sqrt{\pi}$, we get $\lambda = gS_t$: the mean of the Poisson distribution is the density of seedlings. The expected survival probability of a seedling (or the fraction of seedlings surviving), $Z(S_t)$, is obtained by using the survival predictor $z(n)$:

$$Z(S_t) = \sum_{n=0}^{\infty} [p_n \cdot z(n)] = \sum_{n=0}^{\infty} \frac{(gS_t)^n}{n!} e^{-gS_t} z(n). \tag{6.11}$$

The argument of the summation operator, $p_n z(n)$, is the fraction of seedlings having exactly n neighbours and surviving to adulthood.

The fraction of adult plants that had exactly n neighbours when they were seedlings is $p_n z(n)/Z(S_t)$. Therefore, the average number of seeds produced by an adult plant is

$$F(S_t) = \sum_{n=0}^{\infty} \frac{p_n \cdot z(n)}{Z(S_t)} f(n) = \frac{1}{Z(S_t)} \sum_{n=0}^{\infty} \frac{(gS_t)^n}{n!} e^{-gS_t} z(n) f(n), \tag{6.12}$$

where $f(n)$ is an SFP. Substituting (6.11) and (6.12) into the general form (6.9), we get

$$S_{t+1} = gS_t \sum_{n=0}^{\infty} \frac{(gS_t)^n}{n!} e^{-gS_t} z(n) f(n), \tag{6.13}$$

which becomes an explicit expression if the algebraic forms of the performance predictors $z(n)$ and $f(n)$ are known.

Equation (6.13) is the configuration-field model for cases when fecundity is best predicted from the number of seedling neighbours, that is, when $f(n)$ is an SFP. If $f(n)$ is an AFP, $F(S_t)$, the average number of seeds produced per adult plant, can be calculated as

$$F(S_t) = \sum_{n=0}^{\infty} q_n \cdot f(n), \tag{6.14}$$

where q_n is the probability that an adult plant has exactly n adult neighbours. This can be calculated simply by replacing the seedling density gS_t by the adult density $gS_t Z(S_t)$ in the parameter λ of the Poisson distribution, so that

$$q_n = \frac{[gS_t Z(S_t)]^n}{n!} e^{-gS_t Z(S_t)}. \tag{6.15}$$

224   Individual-based neighbourhood models

Substituting this into (6.14), we get

$$F(S_t) = \sum_{n=0}^{\infty} \frac{\left[gS_t Z(S_t)\right]^n}{n!} e^{-gS_t Z(S_t)} \cdot f(n). \tag{6.16}$$

This expression cannot be handled analytically, unless we switch to the simplifying assumption that $Z(S_t)$ is a constant. Note from (6.11) that $Z(S_t)$ is independent of seed density S_t only if the survival predictor $z(n)$ is a constant function, that is, if survival is independent of local density. Making this compromise, we set $Z(S_t) = P$, and substitute it into (6.16) and (6.9) to obtain an analytical configuration-field model for the AFP case:

$$S_{t+1} = gS_t P \cdot \sum_{n=0}^{\infty} \frac{\left[gS_t P\right]^n}{n!} e^{-gS_t P} \cdot f(n). \tag{6.17}$$

The equilibrium and stability criteria can be formulated for the general (implicit) form (6.9), but of course these can be actually checked only for more explicit systems like (6.13) or (6.17), in which the performance predictors are specified. The local stability analysis of these models is a routine task with the standard local linearization method. Depending on the actual performance predictors, both (6.13) and (6.17) can admit locally stable or locally unstable equilibria, with the trajectories being either monotone or oscillatory. If the fixed point is periodically unstable, the trajectories can show divergent oscillations, sustained oscillations (limit cycle behaviour) or chaos. Following Pacala and Silander (1985), I will present more detailed examples for (6.17), substituting three different AFPs into $f(n)$.

With the exponential form $f(n) = Qe^{-vn}$ of the AFP, (6.17) becomes

$$\begin{aligned} S_{t+1} &= gS_t P \cdot \sum_{n=0}^{\infty} \frac{\left[gS_t P\right]^n}{n!} e^{-gS_t P} \cdot Qe^{-vn} \\ &= gS_t PQ \cdot e^{-gS_t P} \sum_{n=0}^{\infty} \frac{\left[gS_t P e^{-v}\right]^n}{n!} \\ &= gS_t PQ \cdot e^{-gS_t P\left(1-e^{-v}\right)} \end{aligned} \tag{6.18}$$

This is a formal equivalent of the exponential logistic model (cf. May, 1976), the behaviour of which depends on the lumped parameter gPQ. Specifically, if $\ln(gPQ) < 1$, the single internal equilibrium point is stable with no oscillations; if $1 < \ln(gPQ) < 2$, the trajectories show damped oscillations around the stable fixed point; if $2 < \ln(gPQ)$, the equilibrium is unstable with trajectories oscillating, even chaotically at large enough values of the gPQ parameter. Numerically calculated trajectories of the configuration-field model (6.18) and the results of the corresponding spatial Monte-Carlo (stochastic) simulation with large dispersal distance are strikingly similar

(Figure 6.11), demonstrating that the Poisson term gives a good estimate for within-neighbourhood density, if the seeds disperse far from the mother plant.

With $f(n) = M/(1 + cn)$, the model is analytically intractable for $c \neq 1$, but numerical solutions indicate that the fixed point can be either stable with monotone trajectories (for $c < 1$), or else it is periodically unstable, possibly with a limit cycle around it ($c > 1$). Of course, $M > 1$ is necessary for fecundity to be greater than one in at least the case with no competitors around, that is, if $n = 0$ – otherwise the system has no positive equilibrium. This version of the configuration-field model is again in good agreement

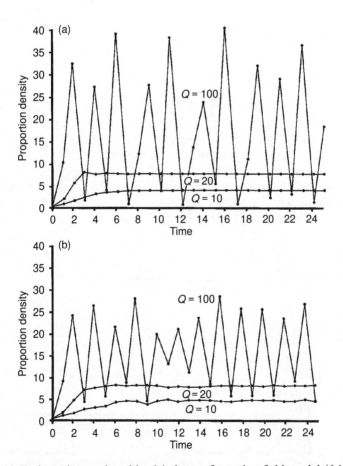

Figure 6.11 Trajectories produced by (a) the configuration-field model (6.18) with the exponential AFP $f(n) = Q \cdot \exp(-vn)$ and (b) the corresponding simulation model with relatively long-range dispersal (dispersal distance was $\frac{4}{3}r$ in all simulations), for different values of Q. Parameters in both cases: $P = 0.2$, $g = 1.0$, $v = 0.2$, $Q = 10, 20, 100$ (after Pacala and Silander (1985)).

226 Individual-based neighbourhood models

with the corresponding Monte-Carlo simulations, if seeds disperse far from their parents (Figure 6.12).

Any realistic fecundity predictor function should be such that it approaches the abscissa gradually as the number of neighbours (local density) increases, but it must not go below zero. Therefore the choice of a simple linear function for the AFP: $f(n) = \gamma - \mu n$ can be justified only for small values of n. Albeit biologically unrealistic, this is still an illuminating case from a theoretical viewpoint, because with a linear AFP, (6.17) reduces to

$$\begin{aligned}S_{t+1} &= gS_tP \cdot \sum_{n=0}^{\infty} \frac{[gS_tP]^n}{n!} e^{-gS_tP} \cdot (\gamma - \mu n) \\ &= \gamma gS_tP - \mu gS_tP \sum_{n=0}^{\infty} \frac{[gS_tP]^n}{n!} e^{-gS_tP} n \\ &= \gamma gS_tP - \mu (gS_tP)^2\end{aligned} \qquad (6.19)$$

(note that the summed expression in the second line of the equation is the mean of a Poisson distribution with parameter gS_tP). Equation (6.19) is a formal equivalent of the discrete logistic map (cf. Section 3.5.1), which also assumes linear density dependence in the growth rate of the population. The formal equivalence of the configuration-field model and the discrete logistic map is not by chance: the assumptions of the two are equivalent, including the complete spatial homogeneity of the environment and the spatial

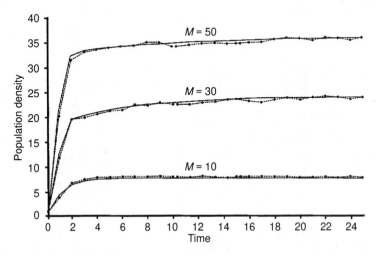

Figure 6.12 Trajectories of (6.17) with the AFP $f(n) = M/(1 + cn)$, and the corresponding simulation model with long-range dispersal. Parameters in both cases: $P = 0.6$, $g = 1.0$, $c = 0.9$, $M = 10, 30, 50$ (after Pacala and Silander (1985)).

stationarity of every relevant demographic process considered. These are in fact the conditions for a configuration-field or a mean-field approximation to be applicable. The mechanism of density-dependence in the configuration-field approach does not differ from that in the non-spatial model, because the density effects are spatially averaged in each: through all possible types of neighbourhoods in the configuration-field model and through the whole habitat in the logistic map. Of course, a similar argument applies – *mutatis mutandis* – to the relation of (6.18) and the exponential logistic model.

The extension of the configuration-field FRN approach to multispecies situations (Pacala, 1986a) is straightforward in principle, albeit it gets too complicated for analytical treatment very soon as s, the number of species, increases. The main idea behind the extension of the single-species model is that the effects of neighbours belonging to different species on the survival and/or the fecundity of the focal individual is different from the effects of conspecifics. Thus the survivorship predictor z_i and the fecundity predictor f_i for species i are functions of multiple arguments n_j, specifying the actual value of the corresponding vital parameter for any composition (n_1, n_2, ..., n_s) of the neighbourhood. The dimensionality of the predictors equals the number of species considered.

In principle at least, the multispecies predictors can be experimentally generated in a way perfectly analogous to the single species case. A very convenient practical property of such empirical multispecies predictors is that they make it unnecessary to assume anything within the model about the way the pairwise competitive effects interact with each other. All the complications of possible synergistic or antagonistic effects among the pairwise competitive interactions are implicit within the predictors themselves. Of course, if one wishes to construct explicit analytical formulae for the predictors, it is indispensable to specify the concrete mathematical form of multispecies interactions. In the simplest possible case without synergisms and antagonisms, pairwise effects are independent of each other, that is, they act multiplicatively or additively. This will be assumed throughout this section for the fecundity predictors. The survival predictors P_i will be kept constant to preserve analytical tractability, just as in the single-species case.

If the neighbourhoods are formed by spatially independent, Poisson distributed individuals of s species, the probability that a neighbourhood of unit area is of composition ($n_1, n_2, \ldots, n_s$) at time t is

$$p_{n_1,n_2,\ldots,n_s} = \prod_{j=1}^{s} \frac{\left(g_j S_{j,t}\right)^{n_j}}{n_j!} \exp\left(-g_j S_{j,t}\right), \tag{6.20}$$

where g_j is the germination success, and $S_{j,t}$ is the average density of species j seeds at time t. It is assumed here that the neighbourhood radius is the same, $1/\sqrt{\pi}$, for all species, but this assumption can be relaxed at the expense

228 Individual-based neighbourhood models

of including new parameters into the model, one for each ordered species pair.

The expected survival probability of an average individual of species i at time t can be calculated using the survival predictor $z_i(n_1, n_2, \ldots, n_s)$ in a way perfectly analogous to the single-species case (cf. (6.11)):

$$Z_i(S_{1,t}, S_{2,t}, \ldots, S_{s,t}) = \sum_{n_1=0}^{\infty} \sum_{n_2=0}^{\infty} \cdots \sum_{n_s=0}^{\infty} \left[P_{n_1, n_2, \ldots, n_s} z_i(n_1, n_2, \ldots, n_s) \right]. \tag{6.21}$$

The expected number of seeds produced by an average individual of species i is

$$F_i(S_{1,t}, S_{2,t}, \ldots, S_{s,t}) = \frac{1}{Z_i(S_{1,t}, S_{2,t}, \ldots, S_{s,t})}$$

$$\sum_{n_1=0}^{\infty} \sum_{n_2=0}^{\infty} \cdots \sum_{n_s=0}^{\infty} \left[P_{n_1, n_2, \ldots, n_s} z_i(n_1, n_2, \ldots, n_s) f_i(n_1, n_2, \ldots, n_s) \right] \tag{6.22}$$

if the fecundity predictor $f_i(n_1, n_2, \ldots, n_s)$ is an SFP (cf. (6.12)). If $f_i(n_1, n_2, \ldots, n_s)$ is the AFP, the expected fecundity of an average individual of species i can be determined analytically only if the average survival probabilities of the seedlings are set constant (independent of local density), that is, if $Z_i(S_{1,t}, S_{2,t}, \ldots, S_{s,t}) = P_i$. Then

$$F_i(S_{1,t}, S_{2,t}, \ldots, S_{s,t}) = \sum_{n_1=0}^{\infty} \sum_{n_2=0}^{\infty} \cdots \sum_{n_s=0}^{\infty} \left[q_{n_1, n_2, \ldots, n_s} f_i(n_1, n_2, \ldots, n_s) \right], \tag{6.23}$$

where

$$q_{n_1, n_2, \ldots, n_s} = \prod_{j=1}^{s} \frac{(g_j P_j S_{jt})^{n_j}}{n_j!} \exp\left[-g_j P_j S_{jt}\right] \tag{6.24}$$

is the proportion of species i adult individuals whose adult neighbourhood composition is $(n_1, n_2, \ldots, n_s)$.

Analogous to the single-species model (cf. (6.9)), the system of recurrence equations

$$S_{i,t+1} = S_{i,t} g_i Z_i(S_{1,t}, S_{2,t}, \ldots, S_{s,t}) F_i(S_{1,t}, S_{2,t}, \ldots, S_{s,t}), \quad i = 1, \ldots, s \tag{6.25}$$

governs the dynamics of the interacting species, whatever specific forms Z_i and F_i take. Substituting

$$f_i(n_1, n_2, \ldots, n_s) = Q_i \exp \sum_{j=1}^{s} (-v_{ij} n_j) \tag{6.26}$$

for the AFP and $Z_i(S_{1,t}, \ldots, S_{s,t}) = P_i$ for survival, (6.25) becomes

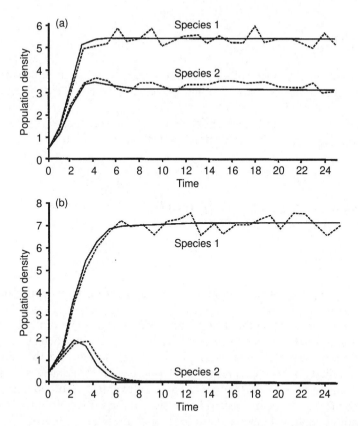

Figure 6.13 Trajectories of the two-species version of (6.27) (exponential AFP) with parameters (a) in the stable coexistence ($c_{12} = c_{21} = 0.1$) and (b) in the competitive exclusion ($c_{12} = c_{21} = 0.3$) domain, and those for the corresponding simulation model with long-range dispersal. Other parameters: $M_1 = 35$, $M_2 = 30$, $c_{11} = c_{22} = 0.2$, $P_1 = P_2 = 0.2$, $g_1 = g_2 = 0.5$ and dispersal distance = $2r$ in both cases. Solid lines: configuration-field model; dotted lines: simulation (after Pacala (1986a)).

$$S_{i,t+1} = S_{i,t} g_i P_i \sum_{n_1=0}^{\infty} \sum_{n_2=0}^{\infty} \cdots \sum_{n_s=0}^{\infty} \left[q_{n_1,n_2,\ldots,n_s} Q_i \exp \sum_{j=1}^{s} (-v_{ij} n_j) \right], \quad i = 1, \ldots, s.$$

(6.27)

Stability analysis with the standard local linearization method reveals that (6.27) can admit the whole range of dynamical behaviour from asymptotic fixed point stability through stable limit cycles to chaos, depending on the number of species, s, and the lumped species parameters $g_i P_i Q_i$. For the simplest two-species case, (6.27) has four equilibria with none, either or both species persistent. The condition for the coexistent (interior) equilibrium to be locally stable is

$$\frac{1-e^{-v_{11}}}{1-e^{-v_{21}}} > \frac{\ln(g_1 P_1 Q_1)}{\ln(g_2 P_2 Q_2)} > \frac{1-e^{-v_{12}}}{1-e^{-v_{22}}}. \tag{6.28}$$

Computer simulation results are again in very good accordance with the predictions of the configuration-field model, if progeny are dispersed far from the parent (Figure 6.13).

The somewhat unrealistic postulate that the fecundity predictor is a linearly decreasing function of neighbourhood density, that is,

$$f_i(n_1, n_2, \ldots, n_s) = M_i - \sum_{j=1}^{s} c_{ij} n_j \tag{6.29}$$

makes the model (6.25) formally identical with the discrete-time s-species Lotka–Volterra model of competitive interaction:

$$X_{i,t+1} = g_i P_i X_{i,t} \left(M_i - \sum_{j=1}^{s} c_{ij} X_{j,t} \right), \tag{6.30}$$

where $X_{i,t} = g_i P_i S_{i,t}$ is the number of adult species i plants having survived to reproduction in time t.

The most interesting theoretical conclusion is that some of the configuration-field approximations to certain FRN systems can be seen as mechanistic, individual-based derivations of phenomenological models that have long been applied in classical, non-spatial population dynamics. The formal equivalence of some FRN models and, for example, the discrete logistic equation (6.19) or the discrete-time version of the Lotka–Volterra competition model (6.30) is in fact a result of their similarity in two respects: (i) the homogeneity assumption regarding the abiotic environment, the dispersal mechanism and population interactions, and (ii) the linear form of density dependence. In a configuration-field model, the density composition of the **average** neighbourhood is the 'miniaturized' image of the global density average, so that calculations with the average neighbourhood necessarily yield the same results as the corresponding non-spatial models. In all, these FRN results can be used as arguments **in favour** of phenomenological modelling in the theory of population dynamics, saying that the kinetic equations can be reconstructed from elementary, individual-based postulates, given strong enough mixing in the population through relatively long-range dispersal. They can also be used as arguments **against** phenomenological modelling, however, by pointing out that overall homogeneity and linear density dependence are rarely applicable for any real situation, and the latter can even lead to nonsense results in the individual-based setting, if density is not very low.

From a mathematical viewpoint, the configuration-field approximations of multispecies FRN systems might be formally tractable in specific cases, which is a valuable property. Even these models suffer from problems similar to those of most classical approaches, however: they become at best

Distance models 231

impractical, and in most cases even impossible to treat analytically as the number of species exceeds three.

From the practical aspect, a great advantage of the FRN approach is that its parameters are relatively easily obtained in experiments lasting for a single generation. This is almost never possible for non-spatial models, for at least two reasons. One is that the parameters of classical models, e.g., the intrinsic growth rate of the population or the carrying capacity of the environment, combine far too many different internal (species specific) and external (milieu specific) factors in an immensely complicated and cryptic manner, which cannot be easily controlled and/or measured by the experimenter. The other problem is that even if it is possible, any proper parameter estimation requires a long enough time series to be recorded for a phenomenological model – but when one measures, for example, the intrinsic growth rate of a population in order to use it as the parameter of a discrete-time model, one can get a single piece of data in every generation.

In conclusion, the fixed radius neighbourhood method seems to be a very useful tool for interfacing population dynamical theory and the field (or laboratory) practice of plant ecology, considering single species or other low-dimensional systems. The practical applicability of FRN modelling has been demonstrated by the authors of the model for single-species cases (Silander and Pacala, 1985; Pacala and Silander, 1987, 1990). Ellison *et al.* (1994) developed a statistical methodology for testing the single-species model and its analytical configuration-field version against field data. It would be interesting to see data from many such single-species and multispecies field experiments: survival and fecundity predictors and dispersal patterns for different species and species pairs. Then the predictive power of the FRN approach could be reliably tested by numerical simulations built directly on experimental data.

6.3.2 Zone of influence models

The fixed radius neighbourhood method is essentially binary in the way the neighbour effects are considered: an individual influences the fate (survival, fecundity) of the focal plant if it falls within the focal plant's FRN, otherwise it does not. Moreover, in the models discussed in the previous section, no state of a neighbour other than its presence within the FRN did matter regarding its impact: once it was within the neighbourhood radius, the focal plant experienced its full effect. Pacala and Silander (1985) point out that the latter assumption is not necessary to make: the actual effect of a neighbour (that is, an individual within the FRN) can be dependent on its geometric relation to the focal plant, like its metric distance or angular position, its age or life cycle stage, or any other i-state descriptor. In principle, these complications can be implemented into the FRN framework, but only at the expense of losing its analytical, numerical and practical tractability altogether. The core of the problem is that increasing the number of relevant

232   Individual-based neighbourhood models

i-states increases the dimensionality of the predictor functions for survival and fecundity, so that they can be neither handled mathematically nor measured experimentally. If, for example, the impact of a neighbour is weighted by a continuous variable, like its Euclidean distance from the focal plant, the appropriate predictor functions should be formally infinite dimensional. The situation becomes not much better if the continuous variable is discretized, because even a few additional dimensions (in this case, one for each distance class for each species) in the predictor state space baffles effectively its numerical, let alone analytical or experimental treatment.

An alternative way to approach neighbourhood interactions is by considering the effects of nearby individuals one by one, instead of calculating with the impact of the whole neighbourhood (possibly many individuals) together. Then the individual effects can be dependent on a larger number of i-state variables, among which the spatial position of the individual is, of course, always included. The effect of a neighbour on the focal plant is usually assumed to decrease with the distance between them increasing, but the actual functional form of distance dependence might be determined from various considerations.

A simplistic yet reasonable example is the assumption that each plant (tiller, ramet) has a circular 'zone of influence' (Hara, 1988) centred on its rooting point, representing the area from which it obtains resources (light, water, nutrients). The focal plant, k, performs best if its ZOI is not overlapped at all by that of others; its relative performance (that is, the value of the affected vital parameter relative to that without interaction), p_{kl}, decreases with A_{kl}, the fraction of its ZOI overlapped by that of neighbour l, increasing. The relative performance function can be a function fitted to experimental or field data, or a conveniently chosen analytical form (Figure 6.14). Simple analytical options for the performance–ZOI overlap function might be the linear $P_{kl} \propto (1 - A_{kl})$, the reciprocal $P_{kl} \propto A_{kl}^{-1}$, or the exponential $P_{kl} \propto \exp(-A_{kl})$. In multispecies systems or age-(stage-)structured models,

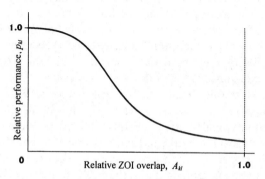

Figure 6.14 Example of the dependence of relative performance (p_{kl}) on relative ZOI overlap (A_{kl}).

the actual value of performance might depend on other *i*-state variables (like the species identity, age, phenophase, etc., of both the neighbour and the focal individual).

For the joint effect P_k of more than one neighbour to calculate, a **collision rule** has to be specified, which determines how multiple ZOI overlaps (multiple neighbour interactions) affect performance. The collision rule can be **simultaneous**, if the neighbours are assumed to act on the focal plant at the same time, or **sequential**, if the interactions with the neighbours take place at different times, in a predetermined or in a random temporal order. With a simultaneous collision rule, the ZOI model becomes theoretically equivalent to a FRN system of potentially very complicated survival and fecundity predictors (cf. Section 6.4.1), whereby the (fixed) neighbourhood radius is equal to the largest ZOI diameter.

The simplest simultaneous collision rule is a multiplicative collision function, assuming that the effect of any single neighbour is independent of the presence of others. Thus the joint effect of the neighbours is the product of the individual effects:

$$P_k = \prod_l p_{kl}. \tag{6.31}$$

More interesting is the case when the collision rule is sequential. Then the interactions occur pairwise and a decision on the fate of the competitors must be made after each such pairwise 'struggle'. Suppose that competition affects survival! If the rule is such that one of the two competitors survive the interaction for certain, we have a simple model of **contest competition** among plants, in which the winner takes all or most of the resources that the contest was for. If the competitive interaction can kill both individuals, the situation is similar to what is called **scramble competition**, where the resources are shared more or less evenly between the competitors, but – at high enough local densities – the shares might be uniformly insufficient for them to survive. It is obvious that if the density is very high everywhere, the whole population could become extinct with the scramble competition mechanism, but this cannot happen in the contest competition case.

There is practically no other choice but to resort to computer simulation for a ZOI model to implement, albeit a simple enough mean-field version might still turn out to be tractable formally (see, for example, Slatkin and Anderson (1984)). Even computer simulations can be rather slow, however, since A_{kl}, the area of ZOI overlap between two individuals, is complicated and time-consuming to calculate for the general case. Omitting the exact geometric details to avoid the awkward calculations while retaining the essential monotonicity features of the overlap function, it is possible to simplify the system considerably. Czárán and Bartha (1989) constructed an age-structured multispecies simulation model (PATPRO) of plant community dynamics, which assumes a simple linear increase in seedling survival probability as the distance from a neighbour increases.

The PATPRO model is initiated by a random pattern of seeds dispersed within the rectangular, continuously scaled plot. Seeds germinate and develop to seedlings, which start competing with each other immediately after emergence; from the second generation – that is, from the time when there may also be adult plants within the plot – seedlings are also under competition, in a one-sided manner, from their adult neighbours. The competitive effect of a neighbour decreases with its distance from the focal seedling; specifically, the performance of the focal seedling increases according to a linear slope function of distance (Figure 6.15). The parameters $p_{kl,\min}$ and d^0_{klm} of the performance function depend on the species identity k of the focal seedling, and the species l and the age m of the competitor. We used a sequential collision rule with the 'contest' competition mechanism for seedling survival, so that at most one of the competitors could be eliminated in a pairwise interaction. The probabilities for the three possible outcomes of the interaction between individuals A and B (A dies and B survives; B dies and A survives; both survive) were determined from the actual values of the seedling performance functions. The physical growth of plant individuals with age was considered as the spatial extension of their zones of influence; we assumed for simplicity that growth is represented by a linear increase of the d^0_{klm} parameter for a given focal species k and competitor species l, with the age m of the competitor. Thus, older plants can affect more distant neighbours.

Stepping from one age class to the next has a density-independent, species- and age-specific mortality component. Each age class of a certain species has a density-independent fecundity value specified, which is the number of viable seeds produced per plant. Seeds are dispersed isotropically around the mother plant according to a Gaussian dispersal distribution: the distance of a dispersed seed is a random number sampled from a zero-mean Gaussian half-distribution. The dispersal parameter is species-dependent; it

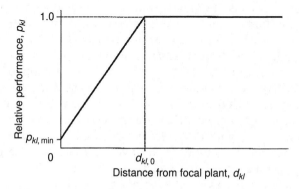

Figure 6.15 The performance p_{kl} of a species k seedling as a function of its distance (D) from a species l seedling in the PATPRO model.

is the standard deviation of the dispersal distribution. Figure 6.16 is a caricature of the pattern-generating mechanism thus defined.

The main purpose of the model has been to approach theoretical population dynamical problems, nevertheless it was also used to produce simulated reference patterns for actual field studies (Czárán and Bartha, 1989). One of the theoretical results (Czárán, 1985) is that communities with cyclically transitive competitive interactions (that is, species a beats b, b beats c, ..., z beats a) become coexistent, and they behave much more 'smoothly' (that is, they show much less violent oscillations of abundance through time) if the interactions are local and dispersal is short-range, as

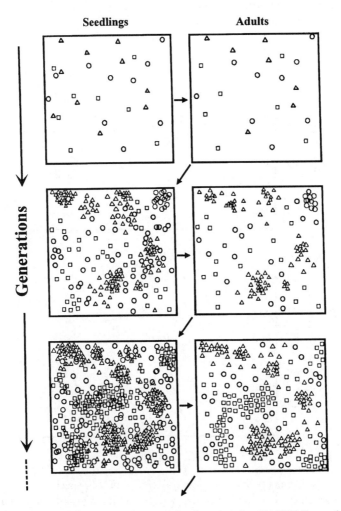

Figure 6.16 The pattern-generating mechanism in the PATPRO model (after Czárán and Bartha (1989)).

compared to the corresponding non-spatial or long-range dispersal system (Figure 6.17). The heuristic reason for this difference is simple: due to short-range dispersal, the populations form nearly monodominant patches, which – due to local competition – 'chase' each other in space, according to the intransitive competitive dominance relations of the community. Thus the overall species abundances are more or less invariant in time, but the community pattern is periodical in space. This is exactly the opposite of the result that we obtain from a non-spatial model of intransitive competitive interactions, which would predict violent periodical oscillations in time, besides complete spatial homogeneity. This result is strongly dependent on both the postulate of the local nature of interaction and that of limited spatial dispersal – violate any of the two and the result will be similar to that of the non-spatial model.

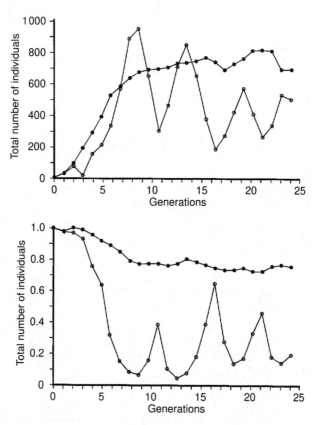

Figure 6.17 Total density and evenness trajectories of a 10-species simulated plant community, with spatially limited Gaussian (filled circles) and with random dispersal (open circles); the competitive relations of the species were cyclically transitive (intransitive) in both simulations (after Czárán (1985)).

Another theoretical problem treated with this model is related to the question of finding community **assembly rules** (Wilson *et al.*, 1987; Wilson, 1989, 1991, 1994) through the analysis of community patterns. On the most general level, assembly rules are considered as the **dynamical constraints** on species assembly processes that determine the **spatial structure** of plant communities. The most often cited such constraint is niche limitation; according to the hypothesis, species with different environmental demands (that is, those with different niches) will rarely co-occur in small enough sampling plots, if the environment is smoothly inhomogeneous, or if it is heterogeneous in a coarse-grained mosaic pattern. According to the expectations based on niche limitation arguments, the spatial pattern of a plant community can be used as an indicator of the rules of the community assembly process. The basic idea behind the search for such assembly rules is that the ecological mechanisms determining the structure of a community can be extracted by statistical means from the spatiotemporal pattern of the community, that is, from the phenomenological 'trace' of those mechanisms. As most dynamical constraints are interpreted as mechanisms of niche limitation in this context, assembly rules are in fact closely related to the competitive relations among the species of the community.

The concrete statistical procedure suggested by Wilson *et al.* (1987) and Wilson (1989, 1991, 1994) for the detection of assembly rules acting in a community is very simple: it is basically the demonstration of a significant variance deficit or variance excess in the number of species per sampling unit, as compared to the variance of the same variable expected in a convenient null model. The null model is derived from a random redistribution of the same number of individuals among the same number of sampling units. The input of the analysis can be either a 'snapshot' map of the community pattern recorded at one single point of time, or a time series of such snapshot maps.

Suspecting that the basic idea of deducing the 'generator' process from the statistical analysis of the generated pattern is seriously mistaken, we simulated a fictitious community dynamical process with the PATPRO model, and tried to extract assembly rules consistent with the simulated mechanism from the time series of the resulting patterns, with the help of multivariate pattern analytical methods, including the detection of variance deficit (or excess) in species numbers per sampling unit (Bartha *et al.*, 1995). One would logically expect that if it is possible to gain assembly rules from a time series of maps, these rules must be identical with, or at least they must be possible to translate to, the dynamical rules 'wired' into the simulation algorithm that produced the maps. This turned out not to be the case, even for extremely simple, thus very characteristic, simulated patterns: the dynamical relations among the populations determined by the parameters of the simulation model did not consistently match those obtained by the statistical analysis of the map series. As for the variance deficit measure in particular, there were simulated cases when strong and characteristic com-

238 Individual-based neighbourhood models

petition relations were masked by the spatial constraints of the pattern-generating mechanism, and in other cases, large variance deficits could be detected without any actual dynamical dependence of the species from each other. This result is not very surprising, given that a certain pattern can be usually produced by a variety of different mechanisms, and any single mechanism can produce many different patterns, depending, for example, on the initial and boundary conditions, not to mention the stochasticity of the dynamics. Thus, the spatial pattern of a community contains insufficient information on the assembly process in general (see also Leps (1990) and Leps and Hadincova (1992)), and sample variance deficit deduced from such a community is even less informative in particular. One needs to obtain data

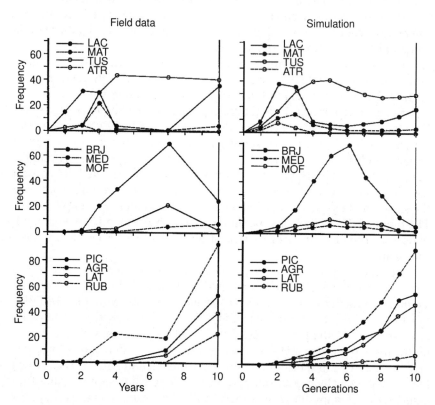

Figure 6.18 Density trajectories from a 10-year data set of a ruderal plant community, compared to a simulated succession process with estimated vital parameters. The three-letter codes correspond to species: LAC = *Lactuca serriola*, MAT = *Matricaria inodora*, TUS = *Tussilago farfara*, ATR = *Atriplex nitens*, BRJ = *Bromus japonicus*, MED = *Medicago lupulina*, MOF = *Melilotus officinalis*, PIC = *Picris hieracioides*, AGR = *Agropyron repens*, LAT = *Lathyrus tuberosus*, RUB = *Rubus caesius* (after Czárán and Bartha (1989)).

Distance models 239

referring directly to the dynamical parameters of population interactions to understand the dynamics of a community. Of course, pattern analysis can indicate certain interactions, but it cannot reliably prove them and still less measure them.

Although the PATPRO model is rather detailed when compared to the usual strategic models of population dynamics, it is not detailed and specific enough for a legitimate comparison of its results to the pattern of complicated ecological entities like an old meadow or a forest. A more realistic model with a lot more parameters would be necessary for that – but increasing the number of parameters inevitably leads to a loss of theoretical insight. It proved useful, however, in simulating simple community dynamical processes such as the invasion of bare soil by ruderal plants. Czárán and Bartha (1989) estimated the dynamical parameters of the most common weed species invading the volcanic tuff dumps created by a strip coal mine in Hungary, and tried to simulate a few years of the dynamics with those parameters. The fits of the simulated abundance trajectories to those of the field case were good (Figure 6.18), suggesting that a simple local competition mechanism might account for the temporal abundance pattern of the community. Of course, the fit of the expected and the observed curve is again not a proof for the validity of the hypothesized competition mechanism in the field case, essentially for the same reason as why the spatial pattern cannot prove, but only indicate, assembly rules. A reliable test of the hypothesis would require correct measurements of the relevant dynamical parameters for each species; beyond doubt, this is difficult even in a simple community as a casual association of ruderal plants.

6.3.3 Ecological field models

A straightforward way for taking the ecological details of the interaction among individual plants into account is to model the spatial distributions of the critical environmental factors as functions of the spatial distribution of plant individuals. The idea is to consider plant individuals not only depending on, but also locally modifying, the environment (water supply, nutrient concentrations, light availability, etc.) around themselves. Individual plants influence the survival, the vitality, the fecundity of their neighbours by reducing the resource levels available to them, or through altering the microclimate or the concentration of allelopathic chemicals to which they might be exposed. In other words instead of simplified, phenomenological assumptions on the distance dependence of interactions between nearby individuals, ecological mechanisms can be directly addressed; the population dynamical consequences of such mechanisms are 'emergent' in this sense.

The most direct formulation of such an ecologically explicit system is due to Wu *et al.* (1985), who published their model under the heading 'ecological field theory' (EFT), emphasizing (to my taste, slightly exaggerating) its similarities with the field theories of physics (electromagnetism, gravitation and nuclear forces). The analogy was based on the general 'effect at a distance' assumption in physical field theories, which the authors adopted in the EFT approach as well. Wu *et al.* (1985) defined the effects of different parts of individual plants (roots, stem and crown) on an originally homogeneous environment separately, in the form of explicit distribution functions for, e.g., soil water availability, light interception or nutrient concentrations around the rooting point of each individual (Figure 6.19). These distributions are defined as the corresponding **ecological fields** generated by a single individual. The joint effect of more individuals on the ecological field is produced by the superimposition of individual distributions. How the individual fields are superimposed depends, of course, on the nature of the ecological factor in question. The joint field for light interception can be quite accurately represented by a simple multiplicative function, but nutrient concentration fields, for example, should be superimposed in a way more complicated than that due to the interaction of limitation effects at different combinations of nutrient concentrations. (Note the analogy of field superimposition rules and the collision rules in other types of neighbourhood models!)

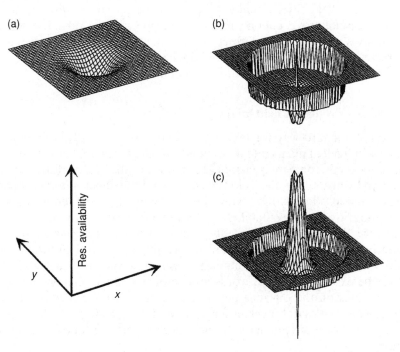

Figure 6.19 Hypothetical examples for (a) soil water availability, (b) light interception and (c) soil nutrient concentration fields around a solitary tree (after Wu *et al.* (1985)).

Distance models 241

The fate of a seed, seedling or adult plant depends on the local values of the ecological fields at the section of space it occupies. With a fine enough temporal resolution, specifying analytical or empirical functions for the dependence of the population dynamical parameters on the field values makes it possible to predict the short-term (within-generation) dynamics of a small community. The basis of the prediction is the so-called **ecological interference potential surface** derived from the field distributions of the different ecological factors and the tolerance properties of the plant species in question (Figure 6.20). This surface specifies the spatial distribution of the probability of survival for a given seedling as a result of the competitive effects of nearby individuals.

It is quite obvious that long-term population dynamical predictions are not sensible with this approach, since the outcome of such spatially small-scale simulations can be expected to depend very sensitively on spatial stochasticity. In other words, the spatial contingencies of propagule dispersion would mask any similarity in the mechanisms of a simulated and an observed process. It is possible with this model, however, at least in principle, to consider different strategies for the exploitation of resource fields, depending on the different physiological means of different species. Clonal plants can, for example, integrate (average) the resource field below a single clone (Oborny and Bartha, 1995; cf. also Section 5.3.7), due to the intraclonal (horizontal) transport of nutrients, assimilates and information,

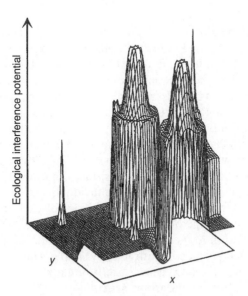

Figure 6.20 Ecological interference field in a fraction of a mixed life-form plant community, derived from the spatial distributions of the ecological factors and species tolerance properties (after Wu *et al.* (1985)).

a property which would be relatively easy to plug into an ecological field model, at the expense of defining additional parameters.

The ecological field approach is completely computer-dependent; no biologically relevant analytical approximations seem possible. Calculating the joint field potential distributions in a community of many individuals would be a time-consuming process even with fast computers, however; therefore the spatial scale at which the EFT approach is applicable and the number of individuals considered in such a model is at least technically limited at the moment. Moreover, the long-term dynamical extension of this basically single-generation model would be possible only with the introduction of many new parameters regarding propagule invasion, dispersion and plant tolerance, each of which should be measured or estimated in any actual field case. This makes the approach difficult to link with other strategic models of theoretical population dynamics. Where it might be very effectively applied is in small-scale, short-term tactic models of actual field cases with practical motivations. Such a problem could be, for example, the determination of optimal sowing patterns in cultivated forests (in monocultures and in mixed cultures alike), if the ecological field distributions generated by the tree species of interest are known. If the sowing pattern is periodical in space – a situation that is quite usual in cultivated forests – it would be sufficient to model a small part of the habitat to assess the dynamics of the whole.

6.4 SUMMARY

Individual-based neighbourhood approaches include tessellation models and distance models. The interacting units in both these types are discrete individuals allocated into continuous space – this is the only feature they share, otherwise the two approaches are very different both in spirit and in formalism. Tessellation models are based on implicit or explicit assumptions concerning habitat space capture by individuals through either individual growth or territorial behaviour. If the dynamical assumptions are implicit, the system is a tessellation assignment model, in which a part of the habitat and the resources within are assigned to every individual, the size of each habitat tile depending on the spatial positions, and possibly also on a few biological parameters, of neighbouring individuals. An attempt is made to predict the fate (survival, vitality, reproduction) of each individual on the basis of the size and shape of the habitat tile it captured, which is assumed to be in direct relation to the share of resources available to the individual. The same approach provides an attractively simple, and also rather general explanation for the $-3/2$ power law of self-thinning in plant monocultures, one of the most robust relationships in ecology.

Temporally explicit tessellations are applied in ecology as models of territoriality and plant growth. In synchronous territory establishment models the individuals arrive at the same time, and they share the habitat more or less evenly by the random tessellation adjustment process. In asynchronous territory models, the individuals arrive sequentially, and each tries to find a sufficiently large tile among the established territories to maintain itself and its offspring. Simple plant growth models are derived from the simultaneous isotropic growth model (SIG) (which leads to the Voronoi tessellation of the habitat space) by relaxing either or both the simultaneity and the homogeneity of the speed of growth assumptions of the SIG process. Tessellation models of long-term population dynamics are yet to be constructed.

Three types of individual-based distance models are applied mainly to plant population and community dynamical problems: fixed radius neighbourhood (FRN) models, zone of influence (ZOI) models and ecological field theory (EFT) approaches. FRN systems define the neighbourhood of the focal individual as the set of individuals within a given distance from its rooting point, the critical distance possibly determined on the basis of biological considerations. The effect of neighbours may or may not be weighted by their actual distance, angular dispersion and biological characteristics. Analytical configuration-field approximations to certain simple FRN systems yield some of the non-spatial models of classical population dynamics. FRN approaches are practically limited to single-species or oligospecies situations, but they can be relatively easily tested against field data.

ZOI models use pairwise distances between individuals to determine the strength of their interaction; specifically, interaction depends on some measure of the overlap between their zones of ecological influence. For modelling the effects of multiple interactions, collision rules need to be defined, but with these kept sufficiently simple, the number of interacting species may be quite large. Details on the spatial growth of ZOIs by plant age can also be relatively easily incorporated, for the sake of improved biological realism, but mean-field or configuration-field approximations are very difficult to obtain for ZOI models.

EFT systems are potentially the most realistic type of distance models in terms of ecological detail. It is possible to consider the effect of the root system and the crown of each individual on the depletion of below- and above-ground resources, and thus on the structure of the ecological interference surface, which feeds back locally to individual growth and reproduction. Because of the detailed mechanistic representation of the interaction between plant individuals, the approach is necessarily small-scale in space and short-term in time, but it can be a useful predictive tool within these spatiotemporal limits even at a practical level.

7 Epilogue

'Pattern and process in the plant community', the pioneering work of Watt (1947), was published just half a century ago. His way of looking at the relation between structural and dynamical aspects of supraindividual organization proved extremely successful in explaining ecological phenomena on all spatiotemporal scales, ranging from population dynamics in ephemeral microhabitats like the calyx of a flower, up to biogeographic processes in an evolutionary time perspective. The diversity of models representing the dynamics of populations and communities in time and space partly corresponds to the diversity of theoretical problems directly emerging from field observations and experiments.

The practical – field – motivation dominates recent modelling efforts in three related topics of applied ecology: forest gap dynamics (Botkin *et al.*, 1972; Shugart, 1984; Leemans and Prentice, 1987; Coffin and Lauenroth, 1990; Prentice and Leemans, 1990; Bushing, 1991; Kawano and Iwasa, 1993; Lauenroth *et al.*, 1993; Iwasa and Kubo, 1995), conservation biology (Soulé, 1986; Lubina and Levin, 1988; Edwards *et al.*, 1994) and landscape ecology (Bormann and Likens, 1979; Forman and Godron, 1986; Gardner *et al.*, 1987; Turner, 1989; Wiens and Milne, 1989; Milne, 1991; di Castri and Hansen, 1992; Johnson *et al.*, 1992). The methodology of these subdisciplines is eclectic: they use all types of spatiotemporal approaches from reaction–diffusion systems to individual-based neighbourhood models – the modelling framework is always chosen so as to fit best to the actual field problem. Different models are in many cases linked with each other and with submodels simulating, for example, water and nutrient cycling (DeAngelis, 1992), to obtain multilevel supermodels (e.g., Lauenroth *et al.*, 1993) intended as tools for long-term predictions on the spatiotemporal dynamics of certain ecosystems. The theoretical value of such a supermodel is limited, as it replaces the immense complexity of the real entity with the incomprehensible complexity of its simulation model. The value of complex ecological supermodels is analogous to that of a flight simulator: it is much simpler, much cheaper and, from a nature conservation point of view, much less offensive to experiment with a computer model than with a real ecosystem. The predictions of a supercomputer model are rarely based on theoretical insight, but they can be of invaluable benefit in the hands of people who design nature reserves or human settlements.

Spatiotemporal modelling in ecology is also motivated by theoretical problems that would not even be possible to formulate without a previous theoretical result. Such a secondary theoretical problem might seem, at first sight, nothing more than an excuse for practising useless mental training – but it is rarely that. Secondary results are very often directly applicable in solving practical problems. To take a simple example: certain initial state problems arising in interacting particle system models induced a huge body of theoretical results on percolations. Outside a percolation approach, no question about the critical level of site connectance would be sensible regarding the spatial spread of epidemics or forest fires, but once the critical values of the percolation parameters are found, they can be translated to the practice of safety measures against the spread of epidemics or fire.

The result of the beneficial convergence of theory and practice in ecology – partly on the basis of spatiotemporal modelling – is an increasing number of studies aiming at a practically useful understanding of critical processes like global extinctions (e.g., Edwards *et al.*, 1994) or the spread of infectious diseases (e.g., Anderson and May, 1991). These, and many other, topics have come a long way from being home affairs of theoretical or experimental ecologists – whether or not we are aware, many aspects of conservation ecology and epidemiology are of immediate concern to each of us living on this planet.

The piers of the bridge connecting research on the spatial structure and the temporal dynamics of biotic communities have been founded. Theoretical and field ecologists are busy building the arcs; parts of the bridge are already passable. Maybe it is not groundless optimism to conclude that the long separation of structural and dynamical, practical and theoretical viewpoints in ecology has come to an end.

Appendix A
The Taylor expansion of univariate and bivariate functions

First we demonstrate the principle of the Taylor expansion on the univariate function $f(x)$, then derive the bivariate version for $p(x, t)$ in its form used throughout Chapter 2.

The method is based on a theorem stating that if a function f of x is continuously differentiable in a neighbourhood of the origin, then it can be replaced with an infinite linear combination of the power functions of x in that neighbourhood so that

$$f(x) = a_0 + a_1 \cdot x + a_2 \cdot x^2 + a_3 \cdot x^3 + \ldots = \sum_{i=0}^{\infty} a_i \cdot x^i. \tag{A.1}$$

The problem is to determine the coefficients a_i. We differentiate f infinitely many times and get the series of derivatives

$$f'(x) = a_1 + 2a_2 x + 3a_3 x^2 + \ldots$$
$$f''(x) = 2a_2 + 2 \cdot 3a_3 x + \ldots$$
$$f'''(x) = 2 \cdot 3 a_3 + \ldots \tag{A.2}$$
$$f^{(n)}(x) = n! a_n + \sum_{j=1}^{\infty} \frac{(n+j)!}{j!} a_{n+j} x^j$$

Note that if the differentials are all evaluated at the origin, that is, at $x = 0$, the term after the summa operator with the powers of x cancels and we are left with the simple relation

$$f^{(n)}(0) = n! a_n, \quad \text{from which} \quad a_n = \frac{f^{(n)}(0)}{n!}. \tag{A.3}$$

Thus the Taylor series expansion of the function f around the origin is

$$f(x) = \sum_{n=0}^{\infty} \frac{f^{(n)}(0)}{n!} \cdot x^n = \sum_{n=0}^{\infty} \frac{1}{n!} \cdot \frac{d^n f(0)}{dx^n} \cdot x^n. \tag{A.4}$$

Appendix A   247

By shifting the origin to the point x, the Taylor expansion of f around x is obtained as

$$f(x+\lambda) = \sum_{n=0}^{\infty} \frac{1}{n!} \cdot \frac{d^n f(x)}{dx^n} \cdot \lambda^n. \tag{A.5}$$

If the function to be expanded is bivariate, like $p(x, t)$ in (2.23), the principle is the same as for the univariate case, but the related algebra is somewhat more involved. If p is a continuously differentiable function in (x, t) in a neighbourhood of the origin $(0, 0)$ including (x, t), then $p(x, t)$ can be expressed as an infinite linear combination of the joint powers of x and t as

$$\begin{aligned}p(x, t) &= a_{00} + (a_{10}x + a_{01}t) + (a_{20}x^2 + 2a_{11}xt + a_{02}t^2) \\ &\quad + (a_{30}x^3 + 3a_{21}x^2t + 3a_{12}xt^2 + a_{03}t^3) + \ldots \\ &= \sum_{n=0}^{\infty} \sum_{i=0}^{\infty} \binom{n}{i} \cdot a_{i,n-i} x^i t^{(n-i)}\end{aligned} \tag{A.6}$$

We wish to determine the $a_{i,n-i}$ coefficients of the series. For this, we proceed in a way completely analogous to that shown for the univariate case, namely, we differentiate $p(x, t)$ as given by (A.6) in the nth order with respect to both x and t in all possible combinations to yield

$$\frac{\partial^n p(x, t)}{\partial x^i \partial t^{n-i}} = n! a_{i,n-i} + \sum_{j=1}^{\infty} \sum_{k=0}^{j} \frac{(n+j)!}{k!(j-k)!} \cdot a_{n+j-k,k} x^{(j-k)} t^k, \quad (i=1,\ldots, n) \tag{A.7}$$

Then we evaluate the derivatives at the origin, that is, at $x = 0$ and $t = 0$, to cancel all terms after the summa operators (which is assured by the fact that either $j - k$ or k is always greater than 0, that is, each term in the summa contains at least one zero factor), and get

$$\frac{\partial^n p(0, 0)}{\partial x^i \partial t^{n-i}} = n! \cdot a_{i,n-i} \quad \text{or} \quad a_{i,n-i} = \frac{1}{n!} \cdot \frac{\partial^n p(0, 0)}{\partial x^i \partial t^{n-i}} \quad (i=1,\ldots, n). \tag{A.8}$$

Substituting (A.8) into (A.6) gives the Taylor expansion of p around the origin:

$$p(x, t) = \sum_{n=0}^{\infty} \frac{1}{n!} \sum_{i=0}^{n} \binom{n}{i} \cdot \frac{\partial^n p(0, 0)}{\partial x^i \partial t^{n-i}} x^i t^{(n-i)}, \quad (i=1,\ldots, n) \tag{A.9}$$

If the origin is shifted into (x, t), we obtain the Taylor expansion of p around (x, t) as

$$p(x+\lambda, t+\tau) = \sum_{n=0}^{\infty} \frac{1}{n!} \sum_{i=0}^{n} \binom{n}{i} \cdot \frac{\partial^n p(x, t)}{\partial x^i \partial t^{n-i}} \lambda^i \tau^{(n-i)}, \quad (i=1,\ldots, n) \tag{A.10}$$

Appendix B
Stability analysis with the local linearization method

Consider an n-dimensional vector function

$$f_i(x_1,\ldots, x_n) \quad (i = 1,\ldots, n)$$

which is partially differentiable at a point $\hat{\mathbf{x}} = \{\hat{x},\ldots, \hat{x}_n\}$! The a_{ij} element of the Jacobian matrix $\mathbf{A}(\hat{\mathbf{x}})$ of this function evaluated at $\hat{\mathbf{x}}$ is the value of $\dfrac{\partial f_i}{\partial x_j}(\hat{\mathbf{x}})$, that is,

$$\mathbf{A}(\hat{\mathbf{x}}) = \begin{bmatrix} \dfrac{\partial f_1}{\partial x_1} & \dfrac{\partial f_1}{\partial x_2} & \cdots & \dfrac{\partial f_1}{\partial x_n} \\ \dfrac{\partial f_2}{\partial x_1} & \dfrac{\partial f_2}{\partial x_2} & \cdots & \dfrac{\partial f_2}{\partial x_n} \\ \vdots & \vdots & \dfrac{\partial f_i}{\partial x_j} & \vdots \\ \dfrac{\partial f_n}{\partial x_1} & \dfrac{\partial f_n}{\partial x_2} & \cdots & \dfrac{\partial f_n}{\partial x_n} \end{bmatrix}(\hat{\mathbf{x}})$$

If f_i are right-hand sides of differential equations in X such that

$$\frac{dx_i}{dt} = f_i(x_1,\ldots, x_n), \quad (i = 1,\ldots, n),$$

the Jacobian matrix evaluated at $\hat{\mathbf{x}}$ contains the coefficients of the corresponding locally linearized system around $\hat{\mathbf{x}}$. a_{ij} is the slope of the tangent of $f_i(\hat{\mathbf{x}})$ in the direction of the x_j axis, which in turn determines how $\dfrac{dx_i}{dt}$, the direction and the speed of change in x_i, depends on a change in x_j. If a_{ij} is negative, for example, then an increase in x_j causes a decrease in x_i and a decrease in x_j causes an increase in x_i.

Suppose that $\hat{\mathbf{x}}$ is an equilibrium point of the original system, so that

$$\frac{dx_i}{dt} = f_i(\hat{x}_1, \ldots, \hat{x}_n) = 0, \quad (i = 1, \ldots, n)$$

The system is locally stable at $\hat{\mathbf{x}}$, if a small perturbation ξ (a change of the equilibrium state from $\hat{\mathbf{x}}$ to $(\hat{\mathbf{x}} + \xi)$) disappears in time, so that the system returns to $\hat{\mathbf{x}}$ asymptotically. This translates to the following statement in matrix algebraic terms. For the fixed point $\hat{\mathbf{x}}$ to be locally stable, the real part of the dominant (or largest real part) eigenvalue λ_{max} of the Jacobian matrix $\mathbf{A}$ of the function $\mathbf{f}$, evaluated at $\hat{\mathbf{x}}$ must be negative (cf. for example Arnold (1981) and Edelstein-Keshet (1988)):

$$\text{Re}\left[\lambda_{max}(\mathbf{A}(\hat{\mathbf{x}}))\right] < 0.$$

Checking this condition for an n-dimensional system requires finding the roots of an n-degree polynomial, which may become impossible even in principle, as the dimensionality of the problem increases above three. A practically still rather laborious, but at least in principle always applicable, equivalent to this condition is the Routh-Hurwitz criterion, for which the reader is referred, for example, to Edelstein-Keshet (1988).

If there are specific constraints, for example, on the sign structure of the Jacobian, sometimes the stability condition above might be reformulated in terms of simple matrix algebraic operations, as shown in Section 3.2.1.

Appendix C
The definition of leading principal minors

An mth-order leading principal minor of a $(k \times l)$ matrix $\mathbf{B}$ is the determinant of the upper left contiguous $(m \times m)$ submatrix of $\mathbf{B}$ ($m \leq k$ and $m \leq l$):

$$\det \begin{bmatrix} b_{11} & \cdots & b_{1m} \\ \vdots & \ddots & \vdots \\ b_{m1} & \cdots & b_{mm} \end{bmatrix}.$$

For a special case of a square matrix $\mathbf{B}$, consider

$$\mathbf{B} = \begin{bmatrix} b_{11} & b_{12} & b_{13} \\ b_{21} & b_{22} & b_{23} \\ b_{31} & b_{32} & b_{33} \end{bmatrix}$$

The first-order leading principal minor of $\mathbf{B}$ is $\det[b_{11}] = b_{11}$. The second-order leading principal minor of $\mathbf{B}$ is

$$\det \begin{bmatrix} b_{11} & b_{12} \\ b_{21} & b_{22} \end{bmatrix} = b_{11}b_{22} - b_{12}b_{21}.$$

The third-order leading principal minor of $\mathbf{B}$ is $\det(\mathbf{B})$.

References

Adler, F.R. and Neuernberger, B. (1994) Persistence in patchy irregular landscapes. *Theor. Popul. Biol.*, **45**, 41–75.

Aguilera, M.O. and Lauenroth, W.K. (1993) Neighbourhood interactions in a natural population of the perennial bunchgrass. *Bouteloua gracilis. Oecologia*, **94**, 595–602.

Allen, J.C. (1975) Mathematical models of species interactions in time and space. *Am. Nat.*, **107**, 319–42.

Allen, L.J.S., Allen, E.J., Kunst, C.R.G. and Sosebee, R.E. (1991) A diffusion model for dispersal of *Opuntia imbricata* (cholla) on rangeland. *J. Ecol.*, **79**, 1123–35.

Anderson, R.M. and May, R.M. (1991) *Infectious Diseases of Humans. Dynamics and Control*, Oxford Science Publications, Oxford.

Andow, D.A., Kareiva, P.M., Levin, S.A. and Okubo, A. (1990) Spread of invading organisms. *Landscape Ecol.*, **4**, 177–88.

Arneodo, A., Coullet, P., Peyraud, J. and Tresser, C. (1982) Strange attractors in Volterra equations for species in competition. *J. Math. Biol.*, **14**, 153–7.

Arnold, V.I. (1981) *Ordinary Differential Equations*, MIT Press, Cambridge, Massachusetts.

Atkinson, W.D. and Shorrocks, B. (1981) Competition on a divided and ephemeral resource: a simulation model. *J. Anim. Ecol.*, **50**, 461–71.

Auld, B.A. and Coote, B.G. (1990) INVADE: towards the simulation of plant spread. *Agric., Ecosyst. Environ.*, **30**, 121–8.

Bailey, V.A., Nicholson, A.J. and Williams, E.J. (1962) Interaction between hosts and parasites when some host individuals are more difficult to find than others. *J. Theor. Biol.*, **3**, 1–18.

Bak, P., Tang, C. and Wiesenfeld, K. (1987) Self-organized criticality: an explanation of 1/f noise. *Phys. Rev. Lett.*, **59**, 381–4.

Bak, P., Tang, C. and Wiesenfeld, K. (1988) Self-organized criticality. *Phys. Rev. A*, **38**, 364–83.

Barlow, G.W. (1974) Hexagonal territories. *Anim. Behav.*, **22**, 876–8.

Bartha, S., Czárán, T. and Oborny, B. (1995) Spatial constraints masking community assembly rules: a simulation study. *Folia Geobot. Phytotax., Praha*, **30**, 471–82.

Beddington, J.R., Free, C.A. and Lawton, J.H. (1975) Dynamic complexity in predator–prey models framed in difference equations. *Nature*, **255**, 58–60.

Begon, M., Harper, J.L. and Townsend, C.R. (1986) *Ecology: Individuals, Populations and Communities*, Blackwell Science, Oxford.

Bell, A.D. (1984) Dynamic morphology: a contribution to plant population ecology, in *Perspectives on Plant Population Ecology* (eds R. Dirzo and J. Sarukhan), Sinauer Associates, Sunderland, Massachusetts.

Ben-Jacob, E., Schocet, O., Tenebaum, A. *et al.* (1994a) Generic modelling of cooperative growth patterns in bacterial colonies. *Nature*, **368**, 46–9.

References

Ben-Jacob, E., Tenenbaum, A., Schochet, O. and Avidan, O. (1994a) Holotransformations of bacterial colonies and genome cybernetics. *Physica A*, **202**, 1–47.

Beretta, E., Solimano, F. and Takeuchi, Y. (1987) Global stability and periodic orbits for two-patch predator–prey diffusion-delay models. *Math. Biosci.*, **85**, 153–83.

Berg, H.C. (1983) *Random Walks in Biology*, Princeton University Press, Princeton.

Bernstein, C., Kacelnik, A. and Krebs, J.R. (1988) Individual decisions and the distribution of predators in a patchy environment. *J. Anim. Ecol.*, **57**, 1007–26.

Bernstein, C., Kacelnik, A. and Krebs, J.R. (1991) Individual decisions and the distribution of predators in a patchy environment. II. The influence of travel costs and the structure of the environment. *J. Anim. Ecol.*, **60**, 205–26.

Boerlijst, M.C. (1994) Selfstructuring: a substrate for evolution. PhD Thesis, Universiteit Utrecht.

Boerlijst, M.C. and Hogeweg, P. (1991) Spiral wave structure in pre-biotic evolution: hypercycles stable against parasites. *Physica D*, **48**, 17–28.

Bormann, F.H. and Likens, G.E. (1979) *Pattern and Process in a Forested Ecosystem*, Springer-Verlag, New York.

Botkin, D.B., Janak, J.F. and Wallis, J.R. (1972) Some ecological consequences of a computer model of forest growth. *J. Ecol.*, **60**, 849–73.

Britton, N.F. (1986) *Reaction–Diffusion Equations and Their Applications to Biology*, Academic Press, New York.

Brown, G.S. (1965) Point density in stems per acre. *NZ For. Res. Notes*, **38**, 1–11.

Buckley, P.A. and Buckley, F.G. (1977) Hexagonal packing of Royal tern nests. *Auk*, **94**, 36–43.

Bülow-Olsen, A., Sackwille Hamilton, N.R. and Hutchings, M.J. (1984) A study of growth form in genets of *Trifolium repens* L. as affected by intra- and interplant contacts. *Oecologia (Berlin)*, **61**, 383–7.

Burkey, T.V. (1989) Extinction in nature reserves: the effect of fragmentation and the importance of migration between reserve fragments. *Oikos*, **55**, 75–81.

Bushing, R.T. (1991) A spatial model of forest dynamics. *Vegetatio*, **92**, 167–92.

Butler, G. and Waltman, P. (1981) Bifurcation from a limit cycle in a two predator-one prey ecosystem modelled on a chemostat. *J. Math. Biol.*, **12**, 295–310.

Cain, M.L. (1990) Models of clonal growth in *Solidago altissima*. *J. Ecol.*, **78**, 27–46.

Callaghan, T.V., Svensson, B.M., Bowman, H. et al. (1990) Models of clonal plant growth based on population dynamics and architecture. *Oikos*, **57**, 257–69.

Cannell, M.G.R., Rothery, P. and Ford, E.D. (1984) Competition within stands of *Picea sithensis* and *Pinus contorta*. *Ann. Bot.*, **53**, 349–62.

Cannings, C. and Cruz Orive, L.M. (1975) On the adjustment of the sex ratio and the gregarious behaviour of animal populations. *J. Theor. Biol.*, **55**, 115–36.

Caswell, H. and Cohen, J.E. (1991a) Communities in patchy environments – a model of disturbance, competition and heterogeneity, in *Ecological Heterogeneity* (eds J. Kolasa and S. Pickett), Springer-Verlag, New York, pp. 97–122.

Caswell, H. and Cohen, J.E. (1991b) Disturbance, interspecific interaction and diversity in metapopulations. *Biol. J. Linn. Soc.*, **42**, 193–218.

Caswell, H. and Etter, R. (1993) Ecological interactions in patchy environments: from patch-occupancy models to cellular automata, in *Patch Dynamics* (eds S.A. Levin and J.H. Steele), Springer-Verlag, Budapest.

Caswell, H. and John, A.M. (1992) From the individual to the population in demographic models, in *Individual-Based Models and Approaches in Ecology* (eds D.L. DeAngelis and L. Gross), Chapman & Hall, New York.
Chaté, H. and Manneville, P. (1988) Continuous and discontinuous transition to spatio-temporal intermittency in two-dimensional coupled map lattices. *Europhys. Lett.*, **6**, 591–5.
Chaté, H. and Manneville, P. (1992) Collective behaviors in coupled map lattices with local and non-local connections. *Chaos*, **2**, 307–13.
Chesson, P.L. (1981) Models for spatially distributed populations: the effect of within-patch variability. *Theor. Popul. Biol.*, **19**, 288–325.
Chesson, P.L. (1985) Coexistence of competitors in spatially and temporally varying environments: a look at the combined effects of different sorts of variability. *Theor. Popul. Biol.*, **28**, 263–87.
Chesson, P.L. and Murdoch, W.W. (1986) Aggregation of risk: Relationships among host–parasitoid models. *Am. Nat.*, **127**, 696–715.
Chewning, W.C. (1975) Migratory effects in predator–prey models. *Math. Biosci.*, **23**, 253–62.
Chow, P.L. and Tam, W.C. (1967) Periodic and travelling wave solutions to Lotka–Volterra equations with diffusion. *Bull. Math. Biol.*, **38**, 643–58.
Coffin, D.P. and Lauenroth, W.K. (1990) A gap dynamics simulation model of succession in a semiarid grassland. *Ecol. Model.*, **49**, 229–66.
Cohen, J.G. (1970) A Markov contingency model for replicated Lotka–Volterra systems near equilibrium. *Am. Nat.*, **104**, 547–59.
Comins, H.N. and Blatt, D.W.E. (1974) Prey–predator models in spatially heterogeneous environments. *J. Theor. Biol.*, **48**, 75–83.
Comins, H.N. and Hassell, M.P. (1979) The dynamics of optimally foraging predators and parasites. *J. Anim. Ecol.*, **48**, 335–51.
Comins, H.N. and Noble, I.R. (1985) Dispersal, variability, and transient niches: species coexistence in a uniformly variable environment. *Am. Nat.*, **126**, 706–23.
Comins, H.N., Hassell, M.P. and May, R.M. (1992) The spatial dynamics of host–parasitoid systems. *J. Anim. Ecol.*, **61**, 735–48.
Connell, J.H. (1978) Diversity in tropical rain forests and coral reefs. *Science*, **199**, 1302–10.
Connell, J.H. and Slatyer, R.O. (1977) Mechanisms of succession in natural communities and their role in community stability and organization. *Am. Nat.*, **111**, 1119–44.
Conroy, M.J., Cohen, Y., James, F.C. *et al.* (1995) Parameter estimation, reliability, and model improvement for spatially explicit models of animal populations. *Ecol. Appl.*, **5**, 17–9.
Cook, R.M. and Hubbard, S.F. (1977) Adaptive searching strategies in insect parasitoids. *J. Anim. Ecol.*, **46**, 115–25.
Cowie, N.R., Watkinson, A. and Sutherland, W.J. (1995) Modelling the growth dynamics of the clonal herb *Anemone nemorosa* in an ancient coppice wood, in *Clonality in Plant Communities* (eds B. Oborny and J. Podani), Proceedings of the 4th International Workshop on Clonal Plants. *Abstracta Botanica*, **19**, 35–50.
Crawley, M.J. and May, R.M. (1987) Population dynamics and plant community structure: competition between annuals and perennials. *J. Theor. Biol.*, **125**, 475–89.

Cronhjort, M.B. and Blomberg, C. (1994) Hypercycles versus parasites in a two dimensional partial differential equations model. *J. Theor. Biol.*, **169**, 31–49.
Crowley, P.H. (1977) Spatially distributed stochasticity and the constancy of ecosystems. *Bull. Math. Biol.*, **39**, 157–65.
Crowley, P.H. (1978) Effective size and the persistence of ecosystems. *Oecologia*, **35**, 185–95.
Crowley, P.H. (1981) Dispersal and the stability of predator–prey interactions. *Am. Nat.*, **118**, 673–701.
Csilling, Á., Jánosi, I.M., Pásztor, G. and Scheuring, I. (1994) Absence of chaos in a self-organized critical coupled map lattice. *Phys. Rev. E*, **50**, 1083–92.
Czárán, T. (1985) A simulation model for generating patterns of sessile populations. *Abstracta Botanica*, **8**, 1–13.
Czárán, T. (1989) Coexistence of competing populations along an environmental gradient: a simulation study. *Coenoses*, **4**, 113–20.
Czárán, T. and Bartha, S. (1989) The effect of spatial pattern on community dynamics: a comparison of simulated and field data. *Vegetatio*, **83**, 229–39.
Czárán, T. and Bartha, S. (1992) Spatio-temporal dynamical models of plant populations and communities. *TREE*, **7**, 38–42.
Czárán, T. and Szathmáry, E. (1997) Coexistence of metabolically co-operating replicators in a cellular automaton: the importance of space without mesoscopic structure, in *Low-dimensional Dynamics of Spatial Ecological Systems* (eds U. Dickman, R. Law and J.A.J. Metz), Springer-Verlag, Berlin (in press).
Davis, T.L.O., Lyne, V. and Jenkins, G.P. (1991) Advection, dispersion and mortality of a patch of southern bluefin tuna larvae *Tunnus maccoyii* in the East Indian Ocean. *Mar. Ecol. Prog. Ser.*, **73**, 33–45.
De Roos, A.M., McCauley, E. and Wilson, W.G. (1991) Mobility versus density limited predator–prey dynamics on different spatial scales. *Proc. R. Soc. Lond. B*, **246**, 117–22.
DeAngelis, D.L. (1992) *Dynamics of Nutrient Cycling and Food Webs*, Chapman & Hall, London.
DeAngelis, D.L. and Gross, L. (eds) (1992) *Individual-Based Models and Approaches in Ecology: Populations, Communities and Ecosystems*, Chapman & Hall, New York.
DeAngelis, D.L., Travis, C.C. and Post, W.M. (1979) Persistence and stability in seed-dispersed species in a patchy environment. *Theor. Popul. Biol.*, **16**, 107–25.
den Boer, P.J. (1968) Spreading of risk and stabilization of animal numbers. *Acta Biotheor. Leiden*, **18**, 165–94.
den Boer, P.J. (1970) Stabilization of animal numbers and the heterogeneity of the environment: the problem of the persistence of sparse populations. *Proc. Adv. Study Inst. Dynamics Numbers Popul.*, Wageningen, pp. 77–9.
Dethier, M.N. (1984) Disturbance and recovery in intertidal pools: maintenance of mosaic patterns. *Ecol. Monogr.*, **54**, 99–118.
Devaney, R.L. (1986) *An Introduction to Chaotic Dynamical Systems*, Benjamin/Cummings, Menlo Park.
di Castri, F. and Hansen, A.D. (eds) (1992) *Landscape Boundaries*, Springer-Verlag, New York.
Diekmann, O. (1978) Thresholds and travelling waves for the geographical spread of infection. *J. Math. Biol.*, **6**, 109–30.

Diekmann, O., Metz, J.A.J. and Sabelis, M.W. (1988) Mathematical models of predator/prey/plant interactions in a patch environment. *Exp. Appl. Acarol.*, **5**, 319–42.

Doak, D.F., Marino, P.C. and Kareiva, P.M. (1993) Spatial scale mediates the influence of habitat fragmentation on dispersal success: implications for conservation. *Theor. Popul. Biol.*, **41**, 315–36.

Dunbar, S. (1983) Travelling wave solutions of diffusive Lotka–Volterra equations. *J. Math. Biol.*, **17**, 11–32.

Dunning, J.B., Stewart, D.J., Danielson, B.J. et al. (1995) Spatially explicit population models: current forms and future uses. *Ecol. Appl.*, **5**, 3–11.

Durrett, R. (1985) *Particle Systems, Random Media, and Large Deviations. AMS Contemporary Mathematics*, Vol. 41, American Mathematical Society.

Durrett, R. (1988) Crabgrass, measles, and gypsy moths: an introduction to interacting particle systems. *The Mathematical Intelligencer*, **10**, 37–47.

Durrett, R. and Levin, S.A. (1994a) The importance of being discrete (and spatial). *Theor. Popul. Biol.*, **46**, 363–94.

Durrett, R. and Levin, S.A. (1994b) Stochastic spatial models: a user's guide to ecological applications. *Phil. Trans. R. Soc. Lond.* B, **343**, 329–50.

Dytham, C. (1994) Habitat destruction and competitive coexistence: a cellular model. *J. Anim. Ecol.*, **63**, 490–1.

Dytham, C. (1995) Competitive coexistence and empty patches in spatially explicit metapopulation models. *J. Anim. Ecol.*, **64**, 145–6.

Edelstein-Keshet, L. (1988) *Mathematical Models in Biology*, Birkhäuser-McGraw-Hill, New York.

Edwards, P.J., May, R.M. and Webb, N.R. (eds) (1994) *Large Scale Ecology and Conservation Biology*, Blackwell, Oxford.

Ellison, A.M., Dixon, P.M. and Ngai, J. (1994) A null model for neighbourhood models of plant competition interactions. *Oikos*, **71**, 225–38

Ermentrout, G.B. and Edelstein-Keshet, L. (1993) Cellular automata approaches in biological modelling. *J. Theor. Biol.*, **160**, 97–133.

Etter, R.J. and Caswell, H. (1993) The advantages of dispersal in a patchy environment: effects of disturbance in a cellular automaton model, in *Reproduction, Larval Biology and Recruitment in the Deep-Sea Benthos* (eds K.J. Eckelbarger and C.M. Young), Columbia University Press.

Evans, G.T. (1980) Diffusive structure: counterexamples to any explanation? *J. Theor. Biol.*, **82**, 313–5.

Fahrig, L. and Merriam, G. (1985) Habitat patch connectivity and population survival. *Ecology*, **66**, 1762–8.

Feller, W. (1952) Parabolic differential equations and associated semigroup transformations. *Ann. Math.*, **55**, 468–518.

Feller, W. (1954) Diffusion processes in one dimension. *Trans. Am. Math. Soc.*, **77**, 1–31.

Feller, W. (1968) *An Introduction to Probability Theory and its Applications*, 3rd edn, Wiley, New York.

Firbank, L.G. and Watkinson, A.R. (1987) On the analysis of competition at the level of the individual plant. *Oecologia (Berlin)*, **71**, 308–17.

Fisher, R.A. (1937) The wave of advance of advantageous genes. *Ann. Eugen. Lond.*, **7**, 355–69.

Fisher, R.A. and Miles, R.E. (1973) The role of spatial pattern in the competition between crop plants and weeds. A theoretical analysis. *Math. Biosci.*, **18**, 335–50.

Fleming, R.A., Marsh, L.M. and Tuckwell, H.C. (1982) Effect of field geometry on the spread of crop disease. *Prot. Ecol.*, **4**, 81–108.

Ford, H. (1987) Investigating the ecological and evolutionary significance of plant growth form using stochastic simulation. *Ann. Bot.*, **59**, 487–94.

Forman, R.T.T. and Godron, M. (1986) *Landscape Ecology*, Wiley, New York.

Free, C.A., Beddington, J.R. and Lawton, J.H. (1977) On the inadequacy of simple models of mutual interference for parasitism and predation. *J. Anim. Ecol.*, **36**, 543–54.

Freedman, H.I. and Takeuchi, Y. (1989a) Global stability and predator dynamics in a model of prey dispersal in a patchy environment. *Non-linear Analysis, Theory, Methods and Applications*, **13**, 993–1002.

Freedman, H.I. and Takeuchi, Y. (1989b) Predator survival versus extinction as a function of dispersal in a predator–prey model with patchy environment. *Applicable Analysis*, **31**, 247–66.

Freedman, H.I. and Waltman, P. (1977) Mathematical analysis of some three-species food-chain models. *Math. Biosci.*, **33**, 257–76.

Fujita, K. (1983) Systems analysis of an acarine predator–prey system II: interactions in discontinuous environment. *Res. Popul. Ecol.*, **25**, 387–99.

Galitsky, V.V. (1990) Dynamic 2D model of plant communities. *Ecol. Model.*, **50**, 95–105.

Gardner, M. (1970) The fantastic combinations of John Conway's new solitaire Game of Life. *Scient. Am.*, April, **223**, 120–3.

Gardner, R.H., Milne, B.T., Turner, M.G. and O'Neill, R.V. (1987) Neutral models for the analysis of broad-scale landscape pattern. *Landscape Ecol.*, **1**, 19–28.

Gerrodette, T. (1981) Dispersal of the solitary coral *Balanophyllia elegans* by demersal planular larva. *Ecology*, **62**, 611–9.

Gilpin, M. and Hanski, I. (eds) Metapopulation dynamics: empirical and theoretical investigations. *Biol. J. Linn. Soc.*, **42**, special issues, no. 1 and 2.

Godfray, H.C.J. (1994) *Parasitoids: Behavioral and Evolutionary Ecology*, Princeton University Press, Princeton.

Godfray, H.C.J. and Grenfell, B.T. (1993) The continuing quest for chaos. *TREE*, **8**, 43–4.

Godfray, H.C.J. and Pacala, S.W. (1992) Aggregation and the population dynamics of parasitoids and predators. *Am. Nat.*, **140**, 30–40.

Godfray, H.C.J., Hassell, M.P. and Holt, R.D. (1994) The population dynamic consequences of phenological asynchrony between parasitoids and their hosts. *J. Anim. Ecol.*, **63**, 1–10.

Gonzalez-Andujar, J.L. and Perry, J.N. (1993) Chaos, metapopulations and dispersal. *Ecol. Model.*, **65**, 255–63.

Grant, P.R. (1968) Polyhedral territories of animals. *Am. Nat.*, **102**, 75–80.

Grassberger, P. (1983) On the critical behaviour of the general epidemic process and dynamical percolation. *Math. Biosci.*, **63**, 157–72.

Green, D.G. (1989) Simulated effects of fire, dispersal and spatial pattern on competition within forest mosaics. *Vegetatio*, **82**, 139–53.

Griffiths, K.J. and Holling, C.S. (1969) A competition submodel for parasites and predators. *Can. Entomol.*, **101**, 907–4.

References 257

Grimm, V., Frank, K., Jeltsch, F. *et al.* (1996) Pattern-oriented modelling in population ecology. *The Science of the Total Environment*, **183**, 151–66.

Gurney, W.S.C. and Nisbet, R.M. (1978) Predator–prey fluctuations in patchy environments. *J. Anim. Ecol.*, **47**, 85–102.

Gyllenberg, M. and Hanski, I. (1992) Single–species metapopulation dynamics: a structured model. *Theor. Popul. Biol.*, **42**, 35–61.

Gyllenberg, M., Söderbacka, G. and Ericsson, S. (1993) Does migration stabilize local population dynamics? Analysis of a discrete metapopulation model. *Math. Biosci.*, **118**, 25–49.

Hamilton, W.D. (1971) Geometry for the selfish herd. *J. Theor. Biol.*, **31**, 295–311.

Hamilton, W.D. and May, R. (1977) Dispersal in stable habitats. *Nature*, **269**, 578–81.

Hanski, I. (1983) Coexistence of competitors in a patchy environment. *Ecology*, **64**, 493–500.

Hanski, I. (1985) Single-species spatial dynamics may contribute to long-term rarity and commonness. *Ecology*, **66**, 335–43.

Hanski, I. (1986) Population dynamics of shrews on small islands accord with the equilibrium model. *Biol. J. Linn. Soc.*, **28**, 23–36.

Hanski, I. (1991a) Metapopulation dynamics: brief history and conceptual domain, in *Metapopulation Dynamics: Empirical and Theoretical Investigations* (eds M. Gilpin and I. Hanski). *Biol. J. Linn. Soc.*, **42**, 3–16.

Hanski, I. (1991b) Single-species metapopulation dynamics: concepts, models and observations, in *Metapopulation Dynamics: Empirical and Theoretical Investigations* (eds M. Gilpin and I. Hanski). *Biol. J. Linn. Soc.*, **42**, 17–38.

Hanski, I. (1994a) A practical model of metapopulation dynamics. *J. Anim. Ecol.*, **63**, 151–62.

Hanski, I. (1994b) Patch-occupancy dynamics in fragmented landscapes. *TREE*, **9**, 131–5.

Hanski, I. and Gyllenberg, M. (1993) Two general metapopulation models and the core-satellite species hypothesis. *Am. Nat.*, **142**, 17–41.

Hanski, I. and Ranta, E. (1983) Coexistence in a patchy environment: three species of Daphnia in rock pools. *J. Anim. Ecol.*, **52**, 263–79.

Hanski, I., Pakkala, T., Kuussaari, M. and Guangchun, L. (1995) Metapopulation persistence of an endangered butterfly in a fragmented landscape. *Oikos*, **72**, 21–8.

Hanson, J.S., Malanson, G.P. and Armstrong, M.P. (1990) Landscape fragmentation and dispersal in a model of riparian forest dynamics. *Ecol. Model.*, **49**, 277–96.

Hara, T. (1985) A model for mortality in a self-thinning plant population. *Ann. Bot.*, **55**, 667–74.

Harper, J.L. (1977) *Population Biology of Plants*, Academic Press, London.

Harrison, S. (1991) Local extinction in a metapopulation context, an empirical evaluation. *Biol. J. Linn. Soc.*, **42**, 73–88.

Harrison, S. (1994) Metapopulations and conservation, in *Large-scale Ecology and Conservation Biology* (eds P.J. Edwards, R.M. May and N.R. Webb), Blackwell, Oxford, pp. 111–28.

Hassell, M.P. (1978) *The Dynamics of Arthropod Predator–Prey Systems*, Princeton University Press, Princeton.

Hassell, M.P. and May, R.M. (1973) Stability in insect host–parasite models. *J. Anim. Ecol.*, **42**, 693–736.

References

Hassell, M.P. and May, R.M. (1974) Aggregation in predators and insect parasites and its effect on stability. *J. Anim. Ecol.*, **43**, 567–94.

Hassell, M.P. and May, R.M. (1988) Spatial heterogeneity and the dynamics of parasitoid–host systems. *Ann. Zool. Fennici*, **25**, 55–61.

Hassell, M.P. and Pacala, S.W. (1990) Heterogeneity and the dynamics of host–parasitoid interactions. *Phil. Trans. R. Soc. Lond. B*, **330**, 203–20.

Hassell, M.P., Comins, H. and May, R.M. (1991) Spatial structure and chaos in insect population dynamics. *Nature*, **353**, 252–8.

Hassell, M.P., Comins, H.N. and May, R.M. (1994) Species coexistence and self-organizing spatial dynamics. *Nature*, **370**, 290–2.

Hassell, M.P., May, R.M., Pacala, S.W. and Chesson, P.L. (1991) The persistence of host–parasitoid associations in patchy environments. I. A general criterion. *Am. Nat.*, **138**, 568–83.

Hastings, A. (1982) Dynamics of a single species in a spatially varying environment: the stabilizing role of high dispersal rates. *J. Math. Biol.*, **16**, 49–55.

Hastings, A. (1991a) Spatial heterogeneity and the stability of predator–prey systems. *Theor. Popul. Biol.*, **12**, 37–48.

Hastings, A. (1991b) Structured models of metapopulation dynamics. *Biol. J. Linn. Soc.*, **42**, 57–71.

Hastings, A. (1993) Complex interactions between dispersal and dynamics: lessons from coupled logistic equations. *Ecology*, **74**, 1362–72.

Hastings, A. and Wolin, C.L. (1989) Within-patch dynamics in a metapopulation. *Ecology*, **70**, 1261–6.

Helland, I.S., Anderbrant, O. and Hoff, J.M. (1989) Modelling bark beetle flight – a review. *Holarct Ecol.*, **12**, 427–31.

Helland, I.S., Hoff, J.M. and Anderbrant, O. (1984) Attraction of bark beetles *Ips Typographus* (Coleoptera, Scolytidae) to a pheromone trap; experimental and mathematical models. *J. Chem. Ecol.*, **10**, 723–2.

Hilborn, R. (1975) The effect of spatial heterogeneity on the persistence of predator–prey interactions. *Theor. Popul. Biol.*, **8**, 346–55.

Hill, A.E. (1990) Pelagic dispersal of Norway lobster (*Nephrops norvegicus*) larvae examined using an advection–diffusion–mortality model. *Mar. Ecol. Prog. Ser.*, **64**, 217–26.

Hirsch, M.W. (1988) Systems of differential equations which are competitive or cooperative: Part III: Competing species. *Nonlinearity*, **1**, 51–71.

Hirsch, M.W. and Smale, S. (1974) *Differential Equations, Dynamical Systems, and Linear Algebra*, Academic Press, New York.

Hofbauer, J. (1990) An index theorem for dissipative semiflows. *Rocky Mountains J. Math.*, **20**, 1017–31.

Hofbauer, J. and Sigmund, K. (1988) *The Theory of Evolution and Dynamical Systems*, Cambridge University Press, Cambridge.

Hofbauer, J. and So, J.W.H. (1994) Multiple limit cycles for three dimensional Lotka–Volterra equations. *Appl. Math. Lett.*, **7**, 65–70.

Hogeweg, P. (1988) Cellular automata as a paradigm for ecological modeling. *Appl. Math. Comput.*, **27**, 81–100.

Holmes, E.E., Lewis, M.A., Banks, J.E. and Veit, R.R. (1994) Partial differential equations in ecology: spatial interactions and population dynamics. *Ecology*, **75**, 17–29.

Holt, R.D., Pacala, S.W., Smith, T.W. and Liu, J. (1995) Linking contemporary vegetation models with spatially explicit animal population models. *Ecol. Appl.*, **5**, 20–7.

Horn, H.S. and MacArthur, R.H. (1972) Competition among fugitive species in a harlequin environment. *Ecology*, **53**, 749–52.

Huffaker, C.B. (1958) Experimental studies on predation: dispersion factors and predator–prey oscillations. *Hilgardia*, **27**, 343–83.

Huston, M. (1979) A general hypothesis of species diversity. *Am. Nat.*, **113**, 81–101.

Huston, M., DeAngelis, D. and Post, W. (1988) New computer models unify ecological theory. *BioScience*, **38**, 682–91.

Hutchings, M.J. and Waite, S. (1985) Cohort behaviour and environmental determination of life histories within a natural population of *Plantago coronopus* L, in *Structure and Functioning of Plant Populations 2* (eds J. Haeck and J.W. Wolendorp), North-Holland, Amsterdam.

Inghe, O. (1989) Genet and ramet survivorship under different mortality regimes – a cellular automata model. *J. Theor. Biol.*, **138**, 257–70.

Ives, A.R. (1988) Aggregation and the coexistence of competitors. *Ann. Zool. Fennica*, **25**, 75–88.

Ives, A.R. (1991) Aggregation and coexistence in a carrion-fly community. *Ecol. Monogr.*, **61**, 75–94.

Ives, A.R. (1992) Density dependent and density independent parasitoid aggregation in model host–parasitoid systems. *Am. Nat.*, **140**, 912–37.

Iwasa, Y. and Kubo, T. (1995) Forest gap dynamics with partially synchronized disturbances and patch age distribution. *Ecol. Model.*, **77**, 257–71.

Jánosi, I.M. (1994) Population dynamics and coupled map lattices, in *Scale Invariance, Interfaces and Non-Equilibrium Dynamics* (ed. A.J. McKane), Plenum, New York.

Jansen, V.A.A. (1994) The dynamics of two diffusively coupled, identical Lotka–Volterra patches, in *Theoretical Aspects of Metapopulation Dynamics*. PhD Thesis, University of Leiden.

Jansen, V.A.A. (1995) Regulation of predator–prey systems through spatial interactions: a possible solution to the paradox of enrichment. *Oikos*, **74**, 384–90.

Jansen, V.A.A. and Sabelis, M.W. (1992) Prey dispersal and predator persistence. *Exp. Appl. Acarol.*, **14**, 215–31.

Jeltsch, F. and Wissel, C. (1994) Modelling dieback phenomena in natural forests. *Ecol. Model.*, **75**, 111–21.

Jeltsch, F., Milton, S.J., Dean, W.R.J. and van Royen, N. (1996) Tree spacing and coexistence in semiarid savannas. *J. Ecol.*, **84** (in press).

Johnson, A.R., Wiens, J.A., Milne, B.T. and Crist, T.O. (1992) Animal movements and population dynamics in heterogeneous landscapes. *Landscape Ecol.*, **7**, 63–75.

Jorné, J. (1974) The effect of ionic migration on oscillations and pattern formation in chemical systems. *J. Theor. Biol.*, **43**, 375–80.

Jorné, J. (1975) Negative ionic cross diffusion coefficients in electrolytic solutions. *J. Theor. Biol.*, **55**, 529–32.

Jorné, J. (1977) The diffusive Lotka–Volterra oscillating system. *J. Theor. Biol.*, **65**, 133–9.

Judson, O.P. (1994) The rise of the individual-based model in ecology. *TREE*, **9**, 9–14.

Juhász-Nagy, P. (1986) The lack, need and tasks of an operative ecology (in Hungarian), Akadémiai Kiadó, Budapest.

Kareiva, P. (1983) Local movement in herbivorous insects: applying a passive diffusion model to mark-recapture field experiments. *Oecologia*, **57**, 322–7.

Kareiva, P. (1990) Population dynamics in spatially complex environments: theory and data. *Phil. Trans. R. Soc. Lond.* B, **330**, 175–90.

Kareiva, P. and Shigesada, N. (1983) Analyzing insect movement as a correlated random walk. *Oecologia*, **56**, 234–8.

Karlson, R.H. (1984) Competition, disturbance and local diversity patterns of substratum-bound clonal organisms: a simulation. *Ecol. Model.*, **23**, 243–55.

Karlson, R.H. and Buss, L.W. (1984) Competition, disturbance and local diversity patterns of substratum-bound clonal organisms: a simulation. *Ecol. Model.*, **23**, 243–55.

Karlson, R.H. and Jackson, J.B.C. (1981) Competitive networks and community structure: a simulation study. *Ecology*, **62**, 670–8.

Kawano, K. and Iwasa, Y. (1993) A lattice structured model for beech forest dynamics: the effect of dwarf bamboo. *Ecol. Model.*, **66**, 261–75.

Kenkel, N.C. (1988) Pattern of self-thinning in jack pine: testing the random mortality hypothesis. *Ecology*, **69**, 1017–24.

Kenkel, N.C. (1990) Spatial competition models for plant populations. *Coenoses*, **5**, 149–58.

Kenkel, N.C. (1993) Modelling Markovian dependence in populations of Aralia nudicaulis. *Ecology*, **74**, 1700–6.

Kenkel, N.C. (1995) Markovian spatial-inhibition models for established clonal populations, in *Clonality in Plant Communities* (eds B. Oborny and J. Podani), Proceedings of the 4th International Workshop on Clonal Plants. *Abstracta Botanica*, **19**, 29–33.

Kenkel, N.C., Hoskins, J.A. and Hoskins, W.D. (1989) Local competition in a naturally established jack pine stand. *Can. J. Bot.*, **67**, 2630–5.

Kerner, E.H. (1959) Further considerations on the statistical mechanics of biological associations. *Bull. Math. Biophys.*, **21**, 217–55.

Kesten, H. (1982) *Percolation Theory for Mathematicians*, Birkhäuser, Boston.

Kierstead, H. and Slobodkin, L.B. (1953) The size of water masses containing plankton bloom. *J. Mar. Res.*, **12**, 141–7.

Kishimoto, K. (1982) The diffusive Lotka–Volterra system with three species can have a stable non-constant equilibrium solution. *J. Math. Biol.*, **16**, 103–12.

Kuang, Y. and Takeuchi, Y. (1994) Predator–prey dynamics in models of prey dispersal in two-patch environments. *Math. Biosci.*, **120**, 77–98.

Lauenroth, W.K., Urban, D.L., Coffin, D.P. et al. (1993) Modelling vegetation structure–ecosystem process interactions across sites and ecosystems. *Ecol. Model.*, **67**, 49–80.

Lawton, J. and Godfray, C. (1990) Case of the diffusing squirrels. *Nature*, **343**, 595–6.

Leemans, R. and Prentice, C. (1987) Description and simulation of tree-layer composition and size distribution in a primeval Picea-Pinus forest. *Vegetatio*, **69**, 147–56.

Leps, J. (1990) Can underlying mechanisms be deduced from observed patterns?, in *Spatial Processes in Plant Communities* (eds F. Krahulec, A.D.Q. Agnew, S. Agnew and J.H. Willems), SBP Publ., Prague, pp. 1–11.

Leps, J. and Hadincova, V. (1992) How reliable are our vegetation analyses? *J. Veget. Sci.*, **3**, 119–24.

Lessels, C.M. (1985) Parasitoid foraging: should parasitism be density-dependent? *J. Anim. Ecol.*, **57**, 27–41.

Levin, S.A. (1974) Dispersion and population interactions. *Am. Nat.*, **108**, 207–28.

Levin, S.A. (1976) Population dynamic models in heterogeneous environments. *A. Rev. Ecol. Syst.*, **7**, 287–310.

Levin, S.A. (1977) A more functional response to predator–prey stability. *Am. Nat.*, **111**, 381–3.

Levin, S.A. (1978) Population models and community structure in heterogeneous environments, in *Studies in Mathematical Biology, Part II: Populations and Communities* (ed. S.A. Levin), MAA Studies in Mathematics, Vol. 16, Mathematical Association of America, Washington, p. 439.

Levin, S.A. (1985a) Population models and community structure in heterogeneous environments, in *Mathematical Ecology. An Introduction. Biomathematics, Vol. 17* (eds T.G. Hallam and S.A. Levin), Springer-Verlag, Berlin, Heidelberg.

Levin, S.A. (1985b) Random walk models of movement and their implications, in *Mathematical Ecology. An Introduction. Biomathematics, Vol. 17* (eds T.G. Hallam and S.A. Levin), Springer-Verlag, Berlin, Heidelberg.

Levin, S.A. and Paine, R.T. (1974) Disturbance, patch formation, and community structure. *Proc. Natl. Acad. Sci. USA*, **71**, 2744–7.

Levin, S.A. and Segel, L.A. (1976) Hypothesis for origin of planktonic patchiness. *Nature*, **259**, 659.

Levins, R. (1966) Strategy of model building in population biology. *Am. Scient.*, **54**, 421–31.

Levins, R. (1968) *Evolution in Changing Environments*, Princeton University Press, Princeton.

Levins, R. (1969) Some demographic and genetic consequences of environmental heterogeneity for biological control. *Bull. Entomol. Soc. Am.*, **15**, 237–40.

Levins, R. (1970) Extinction, in *Some Mathematical Problems in Biology* (ed. M. Gerstenhaber), The American Mathematical Society, Providence, Rhode Island, pp. 77–107.

Levins, R. and Culver, D. (1971) Regional coexistence of species and competition between rare species. *Proc. Natl. Acad. Sci. USA*, **68**, 1246–8.

Liddle, M.J., Budd, C.S.J. and Hutchings, M.J. (1982) Population dynamics and neighbourhood effects in establishing swards of *Festuca rubra*. *Oikos*, **38**, 52–9.

Liddle, M.J., Parlange, J.Y. and Bulow-Olsen, A. (1987) A simple method for measuring diffusion rates and predation of seed on the soil surface. *J. Ecol.*, **75**, 1–8.

Liggett, T. (1985) *Interacting Particle Systems*, Springer-Verlag, New York.

Liu, J. (1993) Discounting initial population sizes for prediction of extinction probabilities in patchy environments. *Ecol. Model.*, **70**, 51–61.

Lomnicki, A. (1992) Population ecology from the individual perspective, *in Individual-based Models and Approaches to Ecology* (eds D. DeAngelis and I.J. Gross), Chapman & Hall, New York, London.

Lomolino, M.V. (1986) Mammalian community structure on islands: the importance of immigration, extinction and interactive effects. *Biol. J. Linn. Soc.*, **28**, 1–21.

Lomolino, M.V., Brown, J.H. and Davis, R. (1989) Island biogeography of montane forest mammals in the American Southwest. *Ecology*, **70**, 180–94.

Lu, Z. and Takeuchi, Y. (1993) Global asymptotic behaviour in single-species discrete diffusion systems. *J. Math. Biol.*, **32**, 67–77.

Lubina, J.A. and Levin, S.A. (1988) The spread of a reinvading species: range expansion in the California sea otter. *Am. Nat.*, **131**, 526–43.

Ludwig, D., Aronson, D.G. and Weinberger, H.F. (1979) Spatial patterning of the spruce budworm. *J. Math. Biol.*, **8**, 217–58.

MacArthur, R.H. and Wilson, E.O. (1967) *The Theory of Island Biogeography*, Princeton University Press, Princeton.

Mack, R. and Harper, J.L. (1977) Interference in dune annuals: spatial pattern and neighbourhood effects. *J. Ecol.*, **65**, 345–63.

MacKay, G. and Jan, N. (1984) Forest fires as critical phenomena. *J. Phys. A*, **17**, L757–L760.

Maley, C.C. and Caswell, H. (1993) Implementing i-state configuration models for population dynamics: an object-oriented programming approach. *Ecol. Model.*, **68**, 75–89.

Mangel, M. and Roitberg, B.D. (1992) Behavioral stabilization of host–parasite population dynamics. *Theor. Popul. Biol.*, **42**, 308–22.

Marsh, L.M. and Jones, R.E. (1988) The form and consequence of random walk movement models. *J. Theor. Biol.*, **133**, 113–31.

Matérn, B. (1960) Spatial variation. *Meddellanden Statens fran Skogsforskningsinstitut Stockholm*, **49**, 1–144.

Matlack, G.R. and Harper, J.L. (1986) Spatial distribution and the performance of individual plants in a natural population of *Silene dioica*. *Oecologia (Berlin)*, **70**, 121–127.

May, R.M. (1976) *Theoretical Ecology: Principles and Applications*, W.B. Saunders Company, Philadelphia.

May, R.M. (1977) Simple mathematical models with very complex dynamics. *Nature*, **261**, 459–67.

May, R.M. (1978) Host–parasitoid systems in patchy environments: A phenomenological model. *J. Anim. Ecol.*, **47**, 833–43.

Maynard Smith, J. (1974) *Models in Ecology*, Cambridge University Press, Cambridge.

McLaughlin, J.F. and Roughgarden, J. (1992) Predation across spatial scales in heterogeneous environments. *Theor. Popul. Biol.*, **41**, 277–99.

McMurtrie, R. (1978) Persistence and stability of single-species and predator–prey systems in spatially heterogeneous environments. *Math. Biosci.*, **39**, 11–51.

Mead, R. (1966) A relationship between individual plant-spacing and yield. *Ann. Botany, NS*, **30**, 301–9.

Mead, R. (1967) A mathematical model for the estimation of inter-plant competition. *Biometrics*, **23**, 189–205.

Mead, R. (1971) Models for interplant competition in irregularly disturbed populations, in *Statistical Ecology. Volume 2, Sampling and Modeling Biological Populations and Population Dynamics* (eds G.P. Patil, E.C. Pielou and W.E. Waters), The Pennsylvania State University Press, Pennsylvania.

Meinhardt, H. (1982) *Models of Biological Pattern Formation*, Academic Press, New York.

Metz, J.A.J and. Diekmann, O. (1986) *The Dynamics of Physiologically Structured Populations*, Springer-Verlag, New York.

Metz, J.A.J. and de Roos, A.M. (1992) The role of physiologically structured population models within a general individual-based modelling perspective, in *Individual-based Models and Approaches to Ecology* (eds D. DeAngelis and I.J. Gross), Chapman & Hall, New York, London.

Milne, B.T. (1991) Lessons from applying fractal models to landscape patterns, in *Quantitative Methods in Landscape Ecology* (eds M.G. Turner and R.H. Gardner), Springer-Verlag, Berlin.

Mimura, M. (1984) Spatial distribution of competing species, in *Mathematical Ecology. Lecture Notes in Biomathematics*, Vol. 54 (eds S.A. Levin and T.G. Hallam), Springer-Verlag.

Mimura, M. and Kawasaki, K. (1980) Spatial segregation in competitive interaction-diffusion equations. *J. Math. Biol.*, **9**, 49–64.

Mimura, M. and Murray, J.D. (1978) On a diffusive prey–predator model which exhibits patchiness. *J. Theor. Biol.*, **75**, 249–62.

Mimura, M. and Nishida, T. (1978) On a certain semilinear parabolic system related to Lotka–Volterra's ecological model. *Publ. Research Inst. Math. Sci. Kyoto Univ.*, **14**, 269–84.

Mithen, R., Harper, J.L. and Weiner, J. (1984) Growth and mortality of individual plants as a function of 'available area'. *Oecologia (Berlin)*, **62**, 57–60.

Moilanen, A. and Hanski, I. (1995) Habitat destruction and coexistence of competitors in a spatially realistic metapopulation model. *J. Anim. Ecol.*, **64**, 141–44.

Munster-Svenson, M. and Nachman, G. (1978) Asynchrony in insect host–parasite interaction and its effect on stability, studied by a simulation model. *J. Anim. Ecol.*, **47**, 159–71.

Murdoch, W.W and Oaten, A. (1975) Predation and population stability. *Adv. Ecol. Res.*, **9**, 1–131.

Murdoch, W.W. and Stewart-Oaten, A. (1989) Aggregation by parasitoids and predators: effects on equilibrium and stability. *Am. Nat.*, **134**, 288–310.

Murdoch, W.W., Briggs, C.J., Nisbet, R.M. *et al.* (1992) Aggregation and stability in metapopulation models. *Am. Nat.*, **140**, 41–58.

Murray, J.D. (1975) Non-existence of wave solutions for the class of reaction-diffusion equations given by the Lotka–Volterra interacting-population equations with diffusion. *J. Theor. Biol.*, **52**, 459–69.

Murray, J.D. (1989) *Mathematical Biology*, Springer-Verlag, New York.

Murray, J.D., Stanley, E.A. and Brown, D.L. (1986) On the spatial spread of rabies among foxes. *Proc. R. Soc. Lond.* B, **229**, 111–51.

Namba, T. (1989) Competition for space in a heterogeneous environment. *J. Math. Biol.*, **27**, 1–16.

Nee, S. and May, R.M. (1992) Dynamics of metapopulations: habitat destruction and competitive coexistence. *J. Anim. Ecol.*, **61**, 37–40.

Nicholson, A.J. and Bailey, V.A. (1935) The balance of animal populations. Part I. *Proc. Zool. Soc. Lond.*, **1935**, 551–98.

Nisbet, R.M. and Gurney, W.S.C. (1982) *Modelling Fluctuating Populations*, John Wiley & Sons, Chichester, New York.

Nisbet, R.M., Briggs, C.J., Gurney, W.S.C. *et al.* (1992) Two-patch population dynamics, in *Patch Dynamics in Freshwater and Marine Ecosystems. Lect. Notes Biomath* (eds S.A. Levin, J.H. Steele and T. Powell), Springer, Berlin, p. 96.

Oborny, B. (1994a) Growth rules in clonal plants and environmental predictability – a simulation study. *J. Ecol.*, **82**, 341–51.

Oborny, B. (1994b) Spacer length in clonal plants and the efficiency of resource capture in heterogeneous environments: a Monte Carlo simulation, in *Plant Clonality: Biology and Diversity* (eds L. Soukupová, C. Marshall, T. Hara, and T. Herben), Opulus Press, Uppsala, pp. 33–52.

Oborny, B. and Bartha, S. (1995) Clonality in plant communities – an overview, in *Clonality in Plant Communities* (eds B. Oborny and J. Podani), Proceedings of the 4th International Workshop on Clonal Plants. *Abstracta Botanica*, **19**, 17–28.

Okabe, A., Boots, B. and Sugihara, K. (1992) *Spatial Tessellations; Concepts and Applications of Voronoi Diagrams*, Wiley & Sons, Chichester.

Okabe, A., Boots, B. and Sugihara, K. (1994) Nearest neighbourhood operations with generalized Voronoi diagrams: a review. *Int. J. G.I.S.*, **8**, 43–71.

Okubo, A. (1978) Horizontal dispersion and critical scales for phytoplankton patches, in *Spatial Pattern in Plankton Communities* (ed. J.H. Steele), Plenum, New York.

Okubo, A. (1980) *Diffusion and Ecological Problems: Mathematical Models. Biomathematics, Vol. 10*, Springer-Verlag, Berlin, Heidelberg, New York.

Okubo, A., Murray, J.D. and Williamson, M. (1989) On the spatial spread of grey squirrels in Britain. *Proc. R. Soc. Lond.* B, **238**, 113–25.

Opdam, P. (1991) Metapopulation theory and habitat fragmentation: a review of holarctic breeding bird studies. *Landscape Ecol.*, **5**, 93–106.

Owens, M.K. and Norton, B.E. (1989) The impact of 'available area' on *Artemisia tridentata* seedling dynamics. *Vegetatio*, **82**, 155–62.

Pacala, S.W. (1986a) Neighborhood models of plant population dynamics. 2 Multispecies models of annuals. *Theor. Popul. Biol.*, **29**, 262–92.

Pacala, S.W. (1986b) Neighborhood models of plant population dynamics. 4 Single-species and multi-species models of annuals with dormant seeds. *Am. Nat.*, **128**, 859–78.

Pacala, S.W. (1987) Neighborhood models of plant population dynamics. 3 Models with spatial heterogeneity in the physical environment. *Theor. Popul. Biol.*, **31**, 359–92.

Pacala, S.W. and Silander, J.A. Jr (1985) Neighborhood models of plant population dynamics. 1 Single-species models of annuals. *Am. Nat.*, **125**, 385–411.

Pacala, S.W. and Silander, J.A. Jr (1987) Neighborhood interference among velvet leaf, *Abutilon theophrasti*, and pigweed, *Amaranthus retroflexus*. *Oikos*, **48**, 217–24.

Pacala, S.W. and Silander, J.A. Jr (1990) Field tests of neighbourhood population dynamic models two annual weed species. *Ecol. Monogr.*, **60**, 113–34.

Pacala, S.W., Hassell, M.P. and May, R.M. (1990) Host–parasitoid associations in patchy environments. *Nature*, **344**, 150–3.

Paine, R.T. and Levin, S.A. (1981) Intertidal landscapes: disturbance and the dynamics of pattern. *Ecol. Monogr.*, 145–78.

Palmer, M.W. (1992) The coexistence of species in fractal landscapes. *Am. Nat.*, **139**, 375–97.

Pearson, K. and Blakeman, J. (1906) Mathematical contributions to the theory of evolution XV: a mathematical theory of random migration. Drapers' Company Research Mem. Biometric Series III, Dept Appl. Math., Univ. College, Univ. London.

Pech, R.P. and McIlroy, J.C. (1990) A model of the velocity of advance of foot and mouth disease in feral pigs. *J. Appl. Ecol.*, **27**, 635–50.

Peltonen, A. and Hanski, I. (1992) Patterns of island occupancy explained by colonization and extinction rates in shrews. *Ecology*, **72**, 1698–708.

Perry, J.N. and Gonzalez-Andujar, J. (1993) Dispersal in a metapopulation neighbourhood model of an annual plant with a seedbank. *J. Ecol.*, **81**, 453–63.

Pickett, S.T.A. and Thompson, J. (1978) Patch dynamics and the design of nature reserves. *Biol. Cons.*, **13**, 27–37.

Pickett, S.T.A. and White, P.S. (1985) *The Ecology of Natural Disturbance and Patch Dynamics*, Academic Press, Orlando.

Pimentel, D., Nigel, W. and Madden, J. (1963) Space–time structure of the environment and the survival of host–parasite systems. *Am. Nat.*, **97**, 141–66.

Possingham, H.P. and Roughgarden, J. (1990) Spatial population dynamics of a marine organism with complex life cycle. *Ecology*, **71**, 973–85.

Prentice, C. and Leemans, R. (1990) Pattern and process and the dynamics of forest structure: a simulation approach. *J. Ecol.*, **78**, 340–55.

Pulliam, H.R. (1988) Sources, sinks, and population regulation. *Am. Nat.*, **132**, 652–61.

Quinn, J.F. and Harrison, S.P. (1988) Effects of habitat fragmentation and isolation on species richness: evidence from biological patterns. *Oecologia*, **75**, 132–40.

Quinn, J.F. and Hastings, A. (1987) Extinction on subdivided habitats. *Cons. Biol.*, **1**, 198–208.

Quinn, J.F., Wolin, C.L. and Judge, M.L. (1989) An experimental analysis of patch size, habitat subdivision, and extinction in a marine intertidal snail. *Cons. Biol.*, **3**, 242–51.

Reddingius, J. and den Boer, P.J. (1970) Simulation experiments illustrating stabilization of animal numbers by spreading of risk. *Oecologia*, **5**, 240–84.

Reeve, J.D. (1988) Environmental variability, migration, and persistence in host–parasitoid systems. *Am. Nat.*, **132**, 810–36.

Reeve, J.D. (1990) Stability, variability, and persistence in host–parasitoid systems. *Ecology*, **71**, 422–6.

Renshaw, E. (1991) *Modelling Biological Populations in Space and Time*. Cambridge University Press, Cambridge.

Ricciardi, L.M. (1985) Stochastic population theory: diffusion processes, in *Mathematical Ecology. An Introduction. Biomathematics, Vol. 17* (eds T.G. Hallam and S.A. Levin), Springer-Verlag, Berlin, Heidelberg.

Ripley, B.D. (1977) Modelling spatial patterns. *J. R. Stat. Soc.* B, **39**, 172–212.

Ripley, B.D. and Kelly, F.P. Markov point processes. *J. Lond. Math. Soc.*, **15**, 188–92.

Roff, D.A. (1974) Spatial heterogeneity and the persistence of populations. *Oecologia*, **15**, 245–58.

Rohani, P. and Miramontes, O. (1995) Host–parasitoid metapopulations: the consequences of parasitoid aggregation on spatial dynamics and searching efficiency. *Proc. R. Soc. Lond.* B, **260**, 335–42.

Rohani, P., Godfray, H.C.J. and Hassell, M.P. (1994) Aggregation and the dynamics of host–parasitoid systems: a discrete-generation model with within-generation redistribution. *Am. Nat.*, **144**, 491–509.

Rosen, G. (1975) Nonexistence of dissipative structure solutions to Volterra many-species models. *J. Math. Phys.*, **16**, 836.

Rosen, G. (1977) Effects of diffusion on the stability of the equilibrium in multispecies ecological systems. *Bull. Math. Biol.*, **39**, 373–83.

Roughgarden, J. and Iwasa, Y. (1986) Dynamics of a metapopulation with space-limited subpopulations. *Theor. Popul. Biol.*, **29**, 235–61.

Sabelis, M.W. and Laane, W.E.M. (1986) Regional dynamics of spider-mite populations that become extinct locally because of food source depletion and predation by Phytoseiid mites (Acarina: Tetranychidae, Phytoseiidae), in *Dynamics of Physiologically Structured Populations* (eds J.A.J. Metz and O. Diekmann), Springer-Verlag, New York.

Sabelis, M.W., Diekmann, O. and Jansen, V.A.A. (1991) Metapopulation persistence despite local extinction: predator–prey patch models of the Lotka–Volterra type. *Biol. J. Linn. Soc.*, **42**, 267–83.

Saupe, D. (1988) Algorithms for random fractals, in *The Science of Fractal Images* (eds H.O. Petigen and D. Saupe), Springer, New York.

Scheuring, I. and Jánosi, I.M. (1996) When two and two make four: a structured population without chaos. *J. Theor. Biol.*, **178**, 89–97.

Scheuring, I., Jánosi, I.M., Csilling, Á. and Pásztor, G. (1993) SOC defeats chaos: A new population dynamical model, in *Proceedings of the IFIP Second International Conference on Fractals in the Natural and Applied Sciences* (ed. M.M. Novak), North Holland, London, Amsterdam.

Segel, L.A. and Jackson, J.L. (1972) Dissipative structure: an explanation and an ecological example. *J. Theor. Biol.*, **37**, 545–59.

Shigesada, N. and Teramoto, E. (1978) A consideration on the theory of environmental density. *Jap. J. Ecol.* **28**, 1–8 (Japanese with an English abstract).

Shigesada, N., Kawasaki, K. and Teramoto, E. (1979) Spatial segregation of interacting species. *J. Theor. Biol.*, **79**, 83–99.

Shorrocks, B. and Swingland, I.R. (eds) (1990) *Living in a Patchy Environment*, Oxford University Press, Oxford.

Shorrocks, B., Atkinson, W. and Charlesworth, P. (1979) Competition on a divided and ephemeral resource. *J. Anim. Ecol.*, **48**, 899–908.

Shugart, H. (1984) *Theory of Forest Dynamics*, Springer-Verlag, New York.

Silander, J.A. and Pacala, S.W. (1985) Neighbourhood predictors of plant performance. *Oecologia*, **66**, 256–63.

Skellam, J.G. (1951) Random dispersal in theoretical populations. *Biometrica*, **38**, 196–218.

Skellam, J.G. (1973) The formulation and interpretation of mathematical models of diffusory processes in population biology, in *The Mathematical Theory of the Dynamics of Biological Populations* (eds M.S. Bartlett and R.W. Hiorns), Academic Press, New York.

Slatkin, M. (1974) Competition and regional coexistence. *Ecology*, **52**, 19–34.

Slatkin, M. and Anderson, D.J. (1984) A model of competition for space. *Ecology*, **65**, 1840–5.

Smale, S. (1976) On the differential equations of species in competition. *J. Math. Biol.*, **3**, 5–7.

Smith, H.L. (1986) Competing subcommunities of mutualists and a generalized Kamke theorem. *SIAM J. Appl. Math.*, **46**, 856–74.

Smith, H.L. (1988) System of ordinary differential equations which generate an order preserving flow: a survey of results. *SIAM Review*, **30**, 87–113.

So, J.W.H. (1979) A note on global stability and bifurcation phenomenon of a Lotka–Volterra food chain. *J. Theor. Biol.*, **80**, 185–7.

Soulé, M.E. (ed.) (1986) *Conservation Biology: the Science of Scarcity and Diversity*, Sinauer Associates, Sunderland, Massachusetts.

Stauffer, D. (1995) *An Introduction to Percolation Theory*, Taylor & Francis, London.

Steele, J.H. (1974) Spatial heterogeneity and population stability. *Nature*, **248**, 83.

Strauss, D.J. (1975) A model for clustering. *Biometrika*, **62**, 467–75.

Strobeck, C. (1973) N-species competition. *Ecology*, **54**, 650–4.

Sutherland, W.J. and Stillman, R.A. (1988) The foraging tactics of plants. *Oikos*, **52**, 239–44.

Takeuchi, Y. (1991) Diffusion mediated persistence in three-species competition models with heteroclinic cycles. *Math. Biosci.*, **106**, 111–28.

Tanemura, M. and Hasegawa, M. (1980) Geometrical models of territory, I. Models for synchronous and asynchronous settlement of territories. *J. Theor. Biol.*, **82**, 477–96.

Tautu, P. (1986) *Stochastic Spatial Processes. Lecture Notes in Mathematics, Vol. 1212*, Springer-Verlag, New York.

Taylor, A. (1988) Large-scale spatial structure and population dynamics in arthropod predator–prey systems. *Ann. Zool. Fenn.*, **25**, 63–74.

Taylor, A.D. (1990) Metapopulation, dispersal, and predator–prey dynamics: an overview. *Ecology*, **71**, 429–33.

Tilman, D., May, R.M., Lehman, C.L. and Nowak, M.A. (1994) Habitat destruction and the extinction debt. *Nature*, **371**, 65–6.

Toffoli, T. and Margolus, N. (1987) *Cellular Automata Machines: a New Environment for Modeling*, MIT Press, Cambridge.

Travis, C.C. and Post, W.M. (1979) Dynamics and comparative statistics of mutualistic communities. *J. Theor. Biol.*, **78**, 553–71.

Turchin, P. (1989) Beyond simple diffusion: models of not-so-simple movement in animals and cells. *Comments Theor. Biol.*, **1**, 65–83.

Turchin, P. (1991) Translating foraging movements in heterogeneous environments into the spatial distribution of foragers. *Ecology*, **72**, 1253–66.

Turing, A.M. (1952) The chemical basis of morphogenesis. *Phil. Trans. R. Soc. Lond.* B, **237**, 37–72.

Turner, M.G. (1989) Landscape ecology: the effects of pattern on process. *Ann. Rev. Ecol. Syst.*, **20**, 171–97.

Turner, M.G., Arthaud, G.J., Engstrom, R.T. *et al.* (1995) Usefulness of spatially explicit population models in land management. *Ecol. Appl.*, **5**, 12–6.

Tychonov, A.N. and Samarski, A.A. (1964) *Partial Differential Equations of Mathematical Physics, Vol. I*, Holden Day, San Francisco.

Tychonov, A.N. and Samarski, A.A. (1967) *Partial Differential Equations of Mathematical Physics, Vol. II*, Holden Day, San Francisco.

Upton, G. and Fingleton, B. (1985) *Spatial Data Analysis by Example, Vol. I, Point Patterns and Qualitative Data*, Wiley, New York.

van Tongeren, O. and Prentice, C.I. (1986) A spatial simulation model for vegetation dynamics. *Vegetatio*, **65**, 163–73.

Vance, R.R. (1984) The effect of dispersal on population stability in one-species, discrete-space population growth models. *Am. Nat.*, **123**, 230–54.

Vandermeer, J. (1993) Loose coupling of predator–prey cycles: entrainment, chaos, and intermittency in the classic MacArthur consumer–resource equations. *Am. Nat.*, **141**, 687–716.

Verboom, J., Lankester, K. and Metz J.A.J. (1991) Linking local and regional dynamics in stochastic metapopulation models. *Biol. J. Linn. Soc.*, **42**, 39–55.

von Neumann, J. (1966) *Theory of Self-Reproducing Automata* (ed. A. Burks), University of Illinois Press, Champaign.

von Niessen, W. and Blumen, A. (1986) Dynamics of forest fires as a directed percolation model. *J. Phys. A*, **19**, L289–L293.

Waldvogel, J. (1986) The period of the Lotka–Volterra system is monotonic. *J. Math. Anal. Appl.*, **114**, 178–84.

Walker, J., Sharpe, P.J.H., Penridge, L.K. and Wu, H. (1989) Ecological field theory: the concept and field tests. *Vegetatio*, **83**, 81–95.

Waller, D.M. (1981) Neighbourhood competition in several violent populations. *Oecologia*, **51**, 116–22.

Watkinson, A.R., Lonsdale, W.M. and Firbank, L.G. (1983) A neighbourhood approach to self-thinning. *Oecologia (Berlin)*, **56**, 381–84.

Watt, A.S. (••) Pattern and process in the plant community. *J. Ecol.*, **35**, 1–22.

Weiner, J. and Conte, P.T. (1981) Dispersal and neighbourhood effects in an annual plant competition model. *Ecol. Model.*, **13**, 131–47.

Whittaker, R.H. and Levin, S.A. (1977) The role of mosaic phenomena in natural communities. *Theor. Popul. Biol.*, **12**, 117–39.

Wiens, J.A. and Milne, B.T. (1989) Scaling of 'landscapes' in landscape ecology, or, landscape ecology from a beetle's perspective. *Landscape Ecol.*, **3**, 87–96.

Wilson, J.B. (1989) A null model for guild proportionality, applied to stratification of a New Zealand temperate rain forest. *Oecologia*, **80**, 263–67.

Wilson, J.B. (1991) Does vegetation science exist? *J. Veg. Sci.*, **2**, 289–90.

Wilson, J.B. (1994) Who makes the assembly rules? *J. Veg. Sci.*, **5**, 275–78.

Wilson, J.B., Gitay, H. and Agnew, A.D.Q. (1987) Does niche limitation exist? *Funct. Ecol.*, **1**, 391–97.

Wilson, W.G., De Roos, A.M. and McCauley, E. (1993) Spatial instabilities within the diffusive Lotka–Volterra system: individual-based simulation results. *Theor. Popul. Biol.*, **43**, 91–127.

Winkler, E. and Schmid, B. (1995) Clonal strategies of herbaceous plant species – a simulation study on population growth and competition, in *Clonality in Plant Communities* (eds B. Oborny and J. Podani), Proceedings of the 4th International Workshop on Clonal Plants. *Abstracta Botanica*, **19**, 17–28.

Wissel, C. and Jeltsch, F. (1993) Modelling pattern formation in ecological systems, in *Interdisciplinary Approaches to Nonlinear Complex Systems* (eds H. Haken and A. Mikhailov), Springer Series in Synergetics, Springer-Verlag, Berlin.

Wolff, W.F. (1994) An-individual-oriented model of a wading bird nesting colony. *Ecol. Model.*, **72**, 75–114.

Wolfram, S. (1986) *Theory and Application of Cellular Automata*, World Scientific, Singapore.

Wu, H., Sharpe, P.J.H., Walker, J. and Penridge, L.K. (1985) Ecological field theory: a spatial analysis of resource interference among plants. *Ecol. Model.*, **29**, 215–43.

Wuensche, A. and Lesser, M. (1992) *The Global Dynamics of Cellular Automata: An Atlas of Basin of Attraction Fields of One-Dimensional Cellular Automata*, Addison-Wesley, Reading, Massachusetts.

Yachi, S., Kawasaki, K., Shigesada, N. and Teramoto, E. (1989) Spatial patterns of propagating waves of fox rabies. *Forma*, **4**, 3–12.

Yoda, K., Kira, T., Ogawa, H., and Hozumi, K. (1963) Self-thinning in over-crowded pure stands under cultivated and natural conditions. *J. Inst. Polytech. Osaka City Univ.* D, **14**, 107–129.

Yodzis, P. (1978) Competition for space and the structure of ecological communities. *Lect. Notes Biomath.*, **25**, 1–191.

Zaninetti, L. (1989) Dynamical Voronoi tessellation. *Astron. Astrophys.*, **224**, 345–50.

Zeigler, B.P. (1977) Persistence and the patchiness of predator–prey systems induced by discrete event population exchange mechanisms. *J. Theor. Biol.*, **67**, 687–713.

Index

Note: page numbers in **bold** refer to figures.

abcentricity 218
  tessellation models 218–19
abundance
  assumption 1
  centre of population 36
  competing species 3
  distribution 4
  drift 62
  long-term dynamics 212
  oscillations 235
  pattern
    stable 75
    temporal xi-xii
  trajectories **238**, 239
adult fecundity predictor (AFP) 221, 224, **225**, 226
  exponential **229**
advection 14, 38–42
  constant rate 39–40, 40–1
  induced by milieu gradient 41–2
  milieu-dependent 42, 55, 56
  term 39
  velocity 41
advective flow 39–40
age structure
  metapopulation 115, 116
  population 234
aggregated interactions 111–12
  models 5, 10, 151
aggregation
  density-dependent habitat choice 90
  host–parasitoid interactions 148–9
  interaction parameters 152
  models
    density dependence 144
    non-spatial 141–4
    spatial interactions 141
    species interactions 140–51

parasitoid
  attacks 149
  spatial distribution 147, 148–9
  prey disperser 92
  spatially undetermined 149–51
  stability effects in predation 92, **94**
  two-patch environment 88–94
Allee effect 55
Arabidopsis thaliana **221**
assembly rules 237
asynchrony
  local 137, 185
  prey emigration 92
attractor structure 163
autocompetition 148–9
autodiffusion, linear 54

bark beetles, release–recapture 41
benefit, mutual 54
biosociology xi, xii–xiii
biotic interactions 109, 141
biotic isolation structure, pairwise 139
birds, nest-building territorial 211–12
boundary
  absorbing 31, **32**, 33, 36
  active 33, 36
  habitat 30
  passive 31
  reflecting 31, **32**, 33, 36, 41
boundary conditions
  cyclic 156
  model 30–1, **32**, 33
Brownian motion 13, 37
  random walk 15
butterfly 139–40
  cabbage 45

carrot 206

Index   271

carrying capacity
  joint 67
  local 189
  single-species 67
carrying simplex 67
catastrophe, local 135, 170
cellular automata (CA) xiv, 7–9, 155, 199
  arena of interaction 202
  attractor structure 163
  deterministic 161
  ecological 164
  emergent properties of pattern 163
  interacting particle systems 156–64
  internal dynamics of complex entities 199–200
  parity rule 160–1
  predator–prey 182–5
  reaction–diffusion systems 200
  self-reproducing system 161
  self-similar patterns 163
  updating rules 159
  see also predator–prey cellular automaton
centroid 211
cereal leaf beetle 45
chaos 98
  dynamics 100
    single-species systems 95–6
chaotic systems 109–10
cichlid fish 210
clonal growth 195
clonal integration 200
  plant competition 195–6, **197**, 198–9
clonal strategy 195–6
clones
  resource field exploitation 241
  territory 216
coenology xi
coexistence
  competitive 55–7, 74–5
  conditions in island populations 69–72
  domain 177
  local asynchrony 185
  parameter combinations 177–8
  spatial pattern 183–4
collision rule 233
colonization
  competing yeast populations 174–5

configuration-field
  approximation 180–2
immigration 140
incidence function models 137
interacting particle systems 173
metacommunity models 134
metapopulation models 114–16, 117
neighbourhood structure 180
per patch rates 120–1, **123**
site state frequency vector 177
state transition route 176
success 115–16
colonization–competition dynamics 174, **175**
community
  assembly rules 237
  dynamics 239
  heterogeneous patterns 185
  matrix 54, 66
  spatial pattern 238
  spatial structure determination 237
competition 54
  contest 233
  dominance hierarchy 134
  extinction 133–4
  facilitation 134, 135
  habitat destruction **173**, 174
  inhibition 135
  initial density of populations 65
  interacting particle systems 173
  interaction intensity 205
  limited in multipatch system 70, 72
  local 236
    stability 72
  Lotka–Volterra process 191
  mean-field model 178, **179**
  migration 133
  neighbourhood-independence 180
  one-sided and temporal refuge effect 174–82
  outcome in IPS model **177**
  scramble 233
  seedling mortality 189–90
  state transition route 176
  for territory 210
  three species model 55
  tolerance 134, 135
  see also autocompetition; plant competition

272 Index

competition–mutualism models, non-spatial 63–4, 66–7
competitive coexistence 55–7, 74–5
competitive dominance relations, intransitive 236
competitive efficiency **191**
competitive exclusion principle 133
competitive interaction
   cyclically transitive 235, **236**
   diffusion 55
   Lotka-Volterra model 230
   non-spatial model of intransitive 236
   per capita growth of species 67
competitive system, s-dimensional 67
competitive transition 176–9
competitive–mutualist
   multispecies–multipatch situation 70, 72
   replicator molecules 171
   scenario 168–9
competitor coexistence through habitat partitioning **56**
configuration entropy, decreasing 163
configuration-field approximation 159, 161–3, 174–82, 178–9, 180–2, 200
configuration-field model 224, 225–7, 229
   density composition 230
   fixed-radius neighbourhood (FRN) models 230
conifers 205
conservation biology 244
   see also nature conservation
constraint functions 6
contact process 198
contest competition 233
core-satellite species
   distribution pattern 130
   hypothesis 123, 125
coupled map lattice (CML) 99–104
   chaos 104–6, **107–8**, 109
   global perturbations 106, 109
   hypercubic lattices 101
   limit cycle 106
   local interaction 185–6
   local population density **107**, **108**
   self-organized criticality 104–6, **107–8**, 109
   short-range dispersion 185–6

spatial parasitoid–host interaction 183
step-by-step delocalization method 103–4
synchronization of lattice 102–3
threshold criterion 106
criticality, self-organized 104–6, **107–8**, 109
cross-diffusion 50, 51
crowding, local 222
crystalline structure 184, 185

deaths due to crowding 166
demography 212–13
density
   flow 25–7
   local rate of change 23
   net efflux 27
density dependence
   Gompertz form 49
   linear 46
density-dependent population interactions 165
destabilizer 54–5
diffusion
   definition 13–14
   density-dependent 37–9, 55
   Fickian 25, 27, 34, 36, 166
   milieu-dependent 55
   random walk approximations 15–21, 22
   range and boundary conditions model 33
diffusion equation 22–32, **33**, 34–8
   density flow 25–7
   flow balance approximation 24–5
   PDF 28
   solutions 34–6
   two- and three-dimensional extensions 34, 35
   two- and three-dimensional spaces 27–9
   weighting factor 35
diffusion model 14–15
   dynamics 33–4
   Lotka–Volterra 49–52
   predator–prey 51
   reactions 42–3
diffusion–advection model 41, 43

diffusive coupling, predator–prey
	patches 82, 84–6
diffusive flow 39–40
diffusive instability, models of
	interacting species 52–5
diffusive systems
	competitive coexistence through
		habitat-partitioning 55–7
	population growth 42–9
	species interactions 49–57
diffusivity
	matrix 51
	virtual 38
Dirac-function 33
direct cascade
	region 102
	scenario 98
Dirichelet tessellation 203, 204, 205
disadvantage, mutual 54
disaster, partial 129
discrete individuality 170–2
	coexistence in pairwise population
		interactions 172
discrete time models 95
discretization, spatial 202
disk equation 81
dispersal/dispersion
	asymmetry and Lotka–Volterra
		two-patch model 86–8
	avalanche/cascade 105–6
	competitive transition 176–9
	connectivity between pairs of
		patches 100
	constant rate 43–5
	distribution 235
	Gaussian 234, 236
	isotropic 234
	limited 179
	mechanism and population
		resilience 76
	parameter 175, 176
	patchy environments 60, 96, 97
	pattern 222
	probability 189
	short-range 235–6
		in reaction–diffusion models
		185–6
	stabilization effects 98–9
	strategies of plants 195

dispersing propagules 171, 196
	patchy environments 59
dispersive coupling, oscillating local
	populations 99–100
dispersivity, competing yeast
	populations 175
dissipative interactions 163–4
distance models xiv, 7–9, 10, 218–42,
	242
	ecological field models 219–20, 239–
	42
	fixed radius neighbourhood (FRN)
		models 219
	zone of influence (ZOI) models 219
distribution, initial 30
disturbance
	abiotic 134
	probability 136
dominance hierarchy 134

ecological field
	generation 240
	models xiv
	theory 239–42
		models 243
ecological interference potential
	surface 241
emigration
	critical abundance level 106
	density-dependent 105
	patch-independent rate for prey and
		predator 88
environment
	inhomogeneities 49
	spatial variation 193
	spatiotemporal pattern 195
environmental gradient, plant
	competition 185–6, 187, 188
environmental grain size 192, 193
environmental heterogeneity 109
environmental inhomogeneity 188
environmental pattern 188, 189, 192
environmental resolution, area of 188,
	192
environmental variation 189, **190**, **192**
epidemics, percolation models 198–9
equilibrium distribution 33
	bimodal 125
ergodicity assumption 1, 2

274　Index

extinction
　competition case 133–4
　conditions for 72
　large subpopulations 126
　local 140
　metacommunity models 130, 133, 134
　metapopulation models 114–15
　metapopulation state 123
　neighbourhood-independence 180
　rate 176
　　competitive transition 176–7
　　per patch 120–1, 122, **123**
　　state space boundary 76
　　threat and habitat fragmentation effects 74

facilitation competition 134, 135
fame dynamics 165
fecundity **191**
　chance 220
　expected 22–3, 220
　individual plant influence on neighbours 239
　predictor 220, 221–2, 223, 228, 230
　function 232
Feigenbaum region 98, 102
Fickian diffusion 26, 34, 36
　model 25
　terms 166
fixed-radius neighbourhood (FRN) models 219, 220–31, 243
　advantages 231
　configuration-field 227
　　model 230
　i-states 231–2
　multispecies 227
　predictive power 231
　single species cases 231
flux density 26
　equilibrium 36
　passive boundary 31
　reaction–diffusion model 50
　reflecting boundary 31
forest fires, percolation models 198–9
forest gap dynamics 244
fractal dimension 189, 190, **191**, **193**
fractal environment 188–95, 200
fugitive species 171

fugitive strategy 174
functional response
　bounded 80–1
　Holling's Type II 81–2, **83**

game step 170
Gause competitive exclusion principle 133
genet 196, 198
germination probability 222
global extinctions 245
Gompertz form for local density dependence 46
Gompertz model, patch-abundance approach 61
growth rate
　constant 43–5
　intrinsic 165
Gyllenberg–Hanski model 124, 125

habitat
　boundary 30
　critical length 46
　critical size 45–6
　　viable 47
　density-dependent choice 90
　destruction and competition effects **173**, 174
　fragmentation 61, 73–4
　heterogeneity effect 123
　minimum length 49
　minimum viable size 49
　n-patch 73
　overlap between species 75
　partitioning 55–7
　patchiness 61
　patchy
　　predation 78–82, **83**, 84–95
　　predator mobility 94–5
　preferential choice 92–3
　sink 46
　size decrease 73–4
　source 46, 73
　see also island biogeography; island populations; patch; patchwork habitat
Hastings–Wolin model 116, 117, 119, 128, 129
Hirsch theorem 67

Index 275

Holling's Type II functional response 81–2, **83**
host, spatial distribution for parasitoids 146
host–parasitoid interactions
  aggregated 141
  population dynamics 88
host–parasitoid models 5
  discrete time 152
hypercyclic replicator dynamics 185
hysteresis 106

i-state
  abundances 6
  configuration 6–7
    models see object-interaction model
  distribution 6
    models see mass-interaction model
  dynamical relevance 5
  space 3, 4
  variables and weighting factors 217
ichneumonids 141
imago 141
immigration 74
  colonization 140
incidence function model 137–40
  spatially semi-explicit system 139
incidence histogram 138
individual-based models 7
infectious disease spread 245
inhibition competition 135
initial condition model 30
initial density pattern 75
initial state configuration 156, 159
instar 141
integrator species **194**, 195, 196, **197**, 198
interacting particle systems (IPSs) xiv, 7–9, 155–6
  arena of interaction 202
  cellular automata 156–64
  coexistence parameter combinations 177–8
  competing metapopulations with habitat destruction 172–4
  competitive transition 176–9
  complexity 163–4

configuration-field
  approximation 159, 161–3, 178–9
cyclic boundary conditions 156
decreasing configuration entropy 163
dispersal 170, 171
dissipative interactions 163–4
dynamical properties 168
ecological 164
epidemics 198–9
forest fires 198–9
frequency-/density-dependent interaction 165, 166–7
game step 170
habitat patches 166–7
initial state configuration 156, 159
internal dynamics of complex entities 199–200
interspecific interactions 167
lattice of sites 156–7
limited dispersal 179
mean-field approximations 159, 161–3
metaphoric models 163
model structure 156–9
mutualist and competitor scenario 169, 170, 171
neighbourhood 167, 169, 170, **171**
  configuration 162
  definition 157–8
next-state function 156, 158–9
parity rule 160–1
payoff matrix 167
percolation models 109
possible states of cells 156, 157
prebiotic evolution model 171–2
reaction–diffusion systems 200
space invariance 157
temporal refuge effect 174–82
two-species predator–prey metacommunity 172
interaction
  area of 202
  coefficient 72
  step 166
interpatch distance 73, 74
intransitive cycle
  facultative mutualism 185
  local interactions 184

276 Index

invasion
  mechanism 183
  neighbourhood 185
invasiveness, predator 184
IPS see interacting particle systems (IPSs)
island biogeography
  colonization by immigration 140
  incidence function models 137–40
  incidence histogram 138
  local extinction 140
  mainland-island model 111
  modelling 140
  species–area relationship 140
island populations 59, 61
  coexistence
    competitive 74–5
    conditions 69–72
  competition and mutualism in dispersing 63–8
  habitat fragmentation effects 73–4
  multispecies multipatch Lotka–Volterra model 68–9
  persistence conditions 69–72
  risk spreading 76–8
  single species
    persistence 73
    resilience 76–8
  spatial pattern 74–5

Jacobian matrix 66, 69, 71, 77, 249
  predation model 79
joint carrying capacity 67

lag *see* time, delay
landscape ecology 244
Lapsana monoculture 208–9, 210
leading principal minors 250
Levins model 114–15, 116, 119
  colonization rate 137
  occupied patches 125
  rescue effect 120
  size structure of patchwork 123
Liapunov test 34
light interception 240
linear dispersal terms 109–10
lobster 45
local asynchrony 109
local competition 236

local interaction 155, 235, 236
  coupled map lattices 185–6
  neighbourhood models 185–6
  reaction–diffusion models 185–6
  subordinated species 179
local linearization 34
local stability
  analysis 224
  competition 72
locust invasion 105
logistic dynamics, synchronous 117–18
logistic growth, Pearl–Verhulst model 1
logistic map, discrete 226
Lotka–Volterra competition process 191
Lotka–Volterra model 1, 3, 49–52
  competitive interaction 230
  host/prey density dependence 143
  multispecies
    multipatch 68–9
    single patch 75
  non-spatial competition–mutualism 63–4, 66–7
  patch-abundance approach 61
  of population interactions 1, 3
  i-state extension 4
  predation xii, 78, **79**, **80**, **82**
    stability properties 82, **83–4**
  spatial homogeneity assumption 14
  time-continuous 142
  two-patch dispersal asymmetry and stability 86–8, **89**
  two-species competition 55

Markov chain
  non-linear 134
    model 178
  pairwise biotic isolation structure 139
Markov matrix, non-linear 181
mass action
  models 1
  population interaction 2
mass-interaction models xiv, 4, 5
mean-field approximation 159, 161–3, **179**, 200, 227
mesoscale pattern 182–5
metacommunity 112–13
  competitive 131–4

Index 277

metacommunity models 170
  asynchrony of local dynamics 137
  colonization 134
  competing predators 135–6
  competing prey populations 135
  continuous time 131–4
  discrete time 134–6
  extinction 130
    rates 133, 134
  global stability 136–7
  immigration 130
  interior equilibria 132
  local catastrophes 135, 170
  local stability analysis 133
  migration competition case 133
  occupancy states 134
    regional 137
  patch occupancy 130
  species coexistence 132
  stable persistence 137
  structure 130–1
  two-species predator–prey 172
metaphoric models 163
metapopulation 5, 111–12
  competing with habitat
    destruction 172–4
  extinction state 123
  interacting 130–1
  metacommunities 112–13
  models 112
metapopulation models 10, 151, 152
  age distribution 116, 124
  bimodal equilibrium distribution 125
  bimodal occupancy pattern 123
  bistability 122–3
    losses 129
  colonization 114–16, **117**
  colonization–extinction
    equilibrium 113–14
  continuous state variable 113
  environment heterogeneity 112–13
  equilibrium occupancy 116–17
    patch 114
  extinction 114–15
    state 124
  frequencies of species
    combinations 113
  habitat heterogeneity effect 123
  interior equilibria 118, **119**

island biogeography 137–40
local dynamics
  asynchrony assumptions 120
  synchronous 117–20
mechanistic model 124–5
multispecies 113
multistate 125–6, **127**, 128–9
occupancy
  high interior equilibrium 122
  levels 121
  patch 124
pairwise biotic isolation
  structure 139
partial disasters 129
per patch extinction rate 122, **123**
rescue effect 123
  due to spatial heterogeneity 120–1
size structure of patchwork 124
stability
  map **129**
  properties of equilibria 128
migrants, patchy environments 59
migration
  competition 135
    case 133
  delayed 95
  habitat fragmentation effects 74
  interpatch 109
  linear 95
  mass 105
  parameters 74
  patch model 166
  state variable 63
milieu factor 42
mites 59–60
monoculture 208–9, 210
Monte-Carlo simulation 224, 226
Moore neighbourhood 158, **160**, 161
  area of environmental resolution
    192
  colonization 174
  predator–prey cellular
    automaton 184
mortality
  due to crowding 166
  seedling 198, 234
mosaic environment 200
mosaic habitat, plant competition for
  space 195–6, **197**, 198–9

mosaic patch
  exploitation 193
  favourable qualification 196
  size 195–6
  variation 195
mosaic pattern, plant competition 188
mouth breeding fish 210
movement
  anisotropic 39
  directed 38
  of individuals 27
  isotropic 27
multipatch system, limited
    competition 70, 72
multiplicative collision function 233
multispecies simulation model,
    age-structured 233
muskrat 45
mutualism
  facultative 185
  obligate 171–2
mutualist and competitor scenario
    168–9

nature conservation
  patch models 61
  see also conservation biology
neighbour, geometric relation to focal
    plant 231, 232
neighbourhood
  configuration 162
  definition in IPS models 157–8
  density composition of average 230
  individual effects of interactions
      232
  interaction 170
  modelling 5–10
  see also Moore neighbourhood;
      Neumann neighbourhood
neighbourhood models xiv, 7
  diversity 9
  fixed radius xiv
  groups 7–9
  individual-based xiv, 202–3
    distance models 218–42
    tessellation models 203–18
  local interaction 185–6
  short-range dispersion 185–6
  site-based 9, 155–6

  see also cellular automata;
      fixed-radius neighbourhood
      (FRN) models; interacting
      particle systems (IPSs)
Neumann neighbourhood **158**, 161, 167
  interacting particle systems **171**
  predator–prey cellular automaton
      183, 184
niche limitation 237
Nicholson–Bailey model 111, 141
  host/prey density dependence 143
  non-spatial 141–4
  parasitoids 141–4
  refuge effect 145
  spatial heterogenity 144–9
  stability maps 147–8
non-spatial models 231
nutrient concentration 240

object-interaction model 7
occupancy
  bimodal pattern 123
  equilibrium 116–17
occupied patch states 125
oceanic islands 59
optimal foraging theory 88, 90
ordinary differential equation (ODE) 4
ordinary differential equation (ODE)
    model
  mutualist and competitor
      scenario 169
  non-spatial 67, 165–6
  payoff matrix 167
  phase portraits **168**

p-state see population state (p-state)
pairwise biotic isolation structure 139
parasite aggregation index 147, **148**
parasitization, probability of
    escape 149
parasitoid–host models, aggregated 111
parasitoids
  aggregation of attacks 149
  attack distribution on host 151
  autocompetition 148–9
  equilibrium pressure 150
  host distribution 150
  non-spatial models 141–4
  pseudo-interference 149, 150, 151

spatial heterogenity models 144–9
spatial patterns 150
stability effects 143
parity rule 160–1
partial differential equations (PDEs) 4
patch
  environmental quality 122
  finite set 139
  interactions 104
  propagule recruitment 62
  size 122
    critical 48
    habitat fragmentation effects 73
  see also habitat, patchy; interpatch
      distance; multipatch system;
      occupied patch states;
      predator–prey patches
patch model
  frequency-/density-dependent
      interaction 165, 166
  mutualistic–competitive payoff 169,
      170
  nature conservation 61
  predatory systems 60
  rules 166
patch-abundance 200
patch-abundance models 4, 60, 61–2,
    151–2
  environment 112
  local dynamics 113
  mathematical formulation 60
  patch-occupancy model comparison
      136–7
  of predation 151
  stability sources 109
  state variable choice 63
patch-occupancy 5, 200
  bistable system 119
  equilibrium 114, 123
  systems 151
patch-occupancy models 111
  environment 112
  local dynamics 113
  metacommunity 130
  patch-abundance model
      comparison 136–7
patchwork habitat
  size structure 124
  spatial spread of population 198

patchy environments 59–61
  asymptotically periodic behaviour 98
  chaos 98
  chaotic dynamics of single-species
      systems 95–6
  dispersion 96, 97
  multiple attractors 98
  patch-abundance approach 61–3
  single-species systems 96–8
  spatial scale splitting 60
  stability region overlap 98
  stabilization effects of dispersion
      98–9
  stable periodic solutions 98, **99**
PATRO model 233–7, 237, 239
pattern generating mechanism 237, 238
payoff matrix 167, 168
Pearl–Verhulst model 61, 96
  of logistic growth 1
percolation
  models 198–9, 200–1
  parameters 245
  theory 199
performance predictor 224
  functions 220–2
performance–zone of influence overlap
    function 232–3
periodic orbits, synchronous
  stability 84–6
  stable disque 85–6, **87**
persistence
  conditions in island populations 69–
      72
  multispecies model 70
  single species 73
perturbation 53, 54
  heterogeneous 76–7
phase transitions, percolation
    theory 199
pheromone trap 41
phytoplankton
  blooms 45
  grazer populations 55
plant
  adult density 223
  performance prediction 206–7
  polygon 206, 209
  population spatial distribution
      modelling 239

280　Index

size variation 195
territory flexibility 219
zone of influence 232
plant competition
　along environmental gradient 185–6, **187**, 188
　clonal integration 195–6, **197**, 198–9
　fractal environment 188–95
　heterogeneity 186
　local segregation **187**, 188
　mosaic habitat space 195–6, **197**, 198–9
　mosaic pattern 188
　response curves 186, 188
Poincaré–Bendixson theorem 67, 68
population
　continuous time invasion process 198
　density-dependent interactions 165
population density
　diffusion equation 23
　frequency 165–6
　state variable 63
　wavefronts 48
population dispersion
　diffusion equation 27
　speed 45
population distribution
　forecasting 33
　homogeneous 43–4
　inhomogeneous 44
　perturbation 34
population dynamics xi
　classical non-spatial models 1–4
　classical theoretical xii
　discrete time models 95
　linked system 212
　model-orientated theory xii-xiii
　spatial constraints 185–6
　threshold-driven phenomena 105
population growth
　logistic 3
　Malthusian 3
　Malthusian model of exponential 44
population growth in diffusive systems
　growth/dispersion
　　constant rate 43–5
　　density-dependent 46–9
　reaction 44
　source term 44

population models, phenomenological 1
population state (p-state) 3
　i-state configuration 6
prebiotic evolution model 171–2
predation
　instar 142
　Lotka–Volterra model xii, 78, **79**, **80**, **82**
　　stability properties 82, **83–4**
　non-spatial Lotka–Volterra system 52
　patch-abundance model 151
　patchy habitats 78–82, **83**, 84–95
　prey density 78
　rate per capita of predator 80
　stability effects of aggregation 92, **94**
predator
　avoidance of conspecifics 89
　competing prey populations 136
　linear migration 95
　mobility 94–5, 184
　per capita predation rate 80
　population efficiency 81
　preferential dispersal 92
　prey density seeking 89
predator–prey cellular automaton
　crystalline structure 184, 185
　invasive mobility 184
　local asynchrony 185
　mesoscale patterns 182–5
　predator invasion 182–3
　predator mobility 184
　site transitions 182
　site-state patterns 18**3**
　spatial pattern of coexistence 183–5
predator–prey diffusion model 51
predator–prey interactions
　aggregated 111
　population dynamics 88
predator–prey patch-abundance models 82
predator–prey patches
　diffusive coupling 82, 84–6
　dispersive coupling 86
　dynamics and regional stability 93–4
　spatial asymmetry 88
　synchronous periodic orbits 84–6
prey
　aggregation impact 91
　Allee effect 55

density and predation 78
density-dependent habitat choice 91, 92, **93**
disperser aggregation 92
dispersion 92
emigration and asynchrony 92
linear migration 95
probability density 20, 24
  estimation (PDE) models 29–31, **32**, 33–4
  function (PDF) 20, **21**, 28–9, **30**
  temporal change of positional **22**
propagules 59, 171, 196
pseudo-interference 149, 150, 151
pseudospatial models 5

ramet 195, 196, 198
  mother 216–17
  spatial spread 215
random walk
  biased 37–9
  continuous time and space 17–21, **22**
  discrete time and space 15–17
  isotropic 26, 34
    in continuous space **19**
    in one dimension **16**
  model and movement of individuals 27
  one-dimensional 28
  probability distribution of spatial position of particle **18**
  speed 20
reaction
  definition 14, 42
  term in Lotka–Volterra-type growth function 51
reaction–diffusion model xiv, 13–15, 165, 166, 167, 168
  critical patch size 48
  cross-diffusion 50
  density-limited population 47
  interactions 50
    local 185–6
  partial differential equation formulation 57
  quadratic growth term 47
  short-range dispersion 185–6
  simple exponential 45
  spiral waves 184, 185

reaction–diffusion system 4, 200
reaction–diffusion–advection systems 57
refuge effect 200
  temporal 174–82
release–recapture experiment 41
replicator dynamics, hypercyclic 185
reproduction, prediction 242
reproductive strategies, plant 195
rescue effect 120–1, 123
resilience
  severely perturbed population **77**, 78
  single-species 76–8
resolution area 188, 192–3, 195
resource
  density 196
  exploitation
    field 241
    territory 214
  intraclonal distribution 196
  limiting 196
  pattern 196, **197**
rhizomes 216
risk spreading 76–8
Routh–Hurwitz criterion 249
ruderal plants **238**, 239

seasonal dynamics of non-spatial systems 109–10
seedling
  density 222, 223
  fecundity predictor 221
  mortality 198, 234
  spatial pattern 207
  survival 233
seeds
  density 224
  viable 222
self-accelerator 54, 55
self-regulation, density-dependent 80
self-regulator 54
self-thinning rule 207–10
sessile population, spatial models 202
Sharkovski's order 98
shrew, incidence histogram **138**
simultaneous isotropic growth (SIG) 204–5, 243
  interfering tiles 217
single species persistence 73

Index

sink habitat 46
sinks 73
site 199–200
   transitions in predator–prey cellular automaton 182
site state
   frequency vector 177
   transition 174
site-based approaches xiv
site-based models 202
site-level rules 200
size-to-resources proportionality assumption 210
soil
   invasion by ruderal plants **238**, 239
   nutrient concentration 240
   water availability 240
source habitat 46
source–sink dynamics 73
sources 73
sowing pattern, optimal 242
space
   exhaustivity 218
   invariance in IPS models 157
spatial constraints, subordinated species 179
spatial discretization 202
spatial distribution 22, 23, 189
   modelling 239
   parasitoids 144–9
   variance 21
spatial diversity, fugitive strategy 174
spatial homogeneity assumption 14
spatial inhibition models 141
spatial interactions, aggregation models 141
spatial models, sessile population 202
spatial pattern xi, 74–5
   biotic environment 202
   community 238
   seedlings 207
spatial position 6
   advection velocity 41
spatial pre-emption 207
spatial randomization 177
spatial scales xi
spatial spread, speed 215–16
spatial structure xi
   determination of community 237

spatial variation 193
spatially astructural–dynamic–analytic approach xi
spatially constrained mechanisms 4
spatiotemporal modelling 155, 244–5
spatiotemporal models 10
spatiotemporal pattern of environment 195
spatiotemporal scale, macroscopic 2
species abundance, vector X 3
species interactions
   aggregation models 140–51
   in diffusive systems 49–57
species–area relationship, log–log linearity 140
spiral waves 184, 185
   rotating 185
splitter species **194**, 195, 196, **197**, 198
spreading populations, percolation models 198–9
stability
   aggregation
     effects 88–90
     of host–parasitoid interactions 148–9
     predator 92, **94**
   Lotka–Volterra two-patch model 86–8, **89**
   neutral 79, 80
   two-patch environment 88–94
stability analysis 69
   with local linearization method 248–9
stabilizer 54–5
stable disque 85–6, **87**
state configuration, initial for IPS 156, 159
state transition rule 156, 159
state variable choice 63
step frequency 37
step length 37
   adjustment 40–1
   anisotropic **40**
step probability
   adjustment 39–40
   anisotropic **40**, 41
steric effects 195
Stirling formula 19
stochastic simulation 224

stochastic systems 185
stochasticity 73–4
stolons 216
strategic models xii
structural–static–statistic discipline xi
supermodels, multilevel 244
survival
  chance 220
  expected 22–3
  individual plant influence on
    neighbours 239
  prediction 242
  predictor 220–1, 222, 224
    function 232
  probability 220, 228–9
symbiosis 54

Taylor expansion of function 24–5
  univariate/bivariate 246–7
temporal abundance patterns xi-xii
temporal change xi
temporal dynamics xi
temporal pre-emption 207
territory
  adcentricity decrease 210–11
  adjustment 210, 211–12
  clone 216
  competition for 210
  establishment 210–12
  fast capture 215
  models and time scale 212
  packing 211, **212**
  polygonal 210
  of resource exploitation 214
tessellation
  additively weighted 215–16, 217, 218
  algorithm 204
  compoundly weighted 216–17
  Direchelet 203, 204, 205
  dynamics 212–13
    multispecies 213–18
  multiplicatively weighted 214–15, 218
  parameters 206–7
tessellation models xiv, 7–9, 10, 203–18, 242
  abcentricity 218–19
  demography 212–13
  dynamic 212, 213

limitations 205
plant performance prediction 206–7
random adjustment 243
self-thinning rule 207–10
temporally explicit 242–3
temporally implicit 206
territory establishment 210–12
weighted 213–18
Thiessen polygon 204
Tilapia mossambica 210
time
  delay 95–6
  dimension in tessellation models 206
  scale in territory models 212
tolerance competition 134, 135
topographical space, resources 218
tuna fish 45
Turing effect 52–5

uniformity assumption 1, 2
updating rules of CA 159

variance deficit/excess, significant 237–8
variation
  adaptive 192
  spatial 193
vector X of species abundance 3
vitality
  individual plant influence on
    neighbours 239
  prediction 242
Voronoi assignment models
  (VVAM) 206
Voronoi diagram 203, 204, 210
  adjustment 211
  normal 215, 216
  regeneration 211
  weighted distances 213–17
Voronoi polygon 204, 205, 206
  centroid 211
  territory packing 211, 212
vote weight 196

waveform, spiral 184
wavefront
  advancing 48
  see also spiral waves
wavelength perturbation, infinite 52
weighted distances 213–17

additively 215–16, 217, 218
  compoundly 216–17
  multiplicatively 214–15, 218
weighting factors, i-state variables 217

yeast populations 174–5

zone of influence (ZOI) 232–3
  collision rule 233
  growth of plant individuals 234
  models xiv, 219, 231–9, 243
zooplanktonic grazer populations 55